"The Marv Tolman science activity books were alway
and were valuable for use with my science method course students."

—Dr. Donald R. Daugs, professor emeritus, Utah State University

"I love the Hands-On Science books by Marv Tolman. They sit right next to my desk within easy reach. The lesson plans are teacher friendly and student friendly. The books are put together in such a way that it is easy to find any subject. From Kindergarten to 6th grade the lessons are set up for all students to use. I know, I have used them in all grades and my students love to be involved in science. I highly recommend these books for any teacher whether you struggle with science or are an expert in the field."

—Marilyn Bulkley, fifth grade teacher, Garfield School District,
Panguitch, Utah

"Our sixth, seventh, and eighth grade students have literally worn out our copies of Dr. Tolman's Hands-On Activities. We have two copies of each activity book and students select them for their concise instructions, excellent diagrams, and easy-to-find materials. Even though the title indicates grades K–6, students and teachers in seventh and eighth grades found the activities very helpful in classroom demonstrations and as a beginning point for science fair problems. I loved the way that the activities are divided into the three different disciplines: physical, earth, and life. It was fast and easy to locate just what you needed using the table of contents organization. There was always an activity for whatever concept my students were learning. Our school will need to purchase the newly revised activity books."

—Rosalee Riddle, science teacher, Red Hills Middle School,
Richfield, Utah and science curriculum coordinator,
Sevier School District, Richfield, Utah

"Before my current position as science teacher educator, I used these activity books extensively in teaching science to my sixth graders. I found them very valuable and helpful in creating lessons that engaged the students in hands-on activities that effectively taught science concepts. Now, as an assistant professor of science education, I continue to use them as I discuss and model hands-on and inquiry-based science. Additionally, I highly recommend them to the pre-service students in my elementary science methods courses, many of whom use them in lesson planning and unit planning for my class as well as during their practicum and student teaching experiences."

—Leigh K. Smith, Ph.D., assistant professor,
Department of Science Education,
Brigham Young University

"I have taught many science classes for our district over the last ten years, and I always rely on your Hands-On books for background information. I have them pulled off of the shelf now because I am using them to come up with a science/math activity for my National Board Certification Program. I know I can trust these books. The information is clearly presented, easy to understand, and the activities always work."

—Mary Selin, second grade teacher, Davis District and
first grade science trainer for the district,
Davis School District, Farmington, Utah

"I have used Marv's science books for several years. I love the experiments. The kids love to try them on their own. They have easy-to-follow directions and all the materials are readily available. I especially like the explanations of how and why. The best thing about these experiments is that they work! I haven't had one fail yet."

—Keetette Turner, kindergarten teacher,
Granite District, Salt Lake City, Utah

"I began working with Marv Tolman ten years ago. I was new to the world of elementary teaching. I had heard of a workshop that helped teachers simplify and pinpoint science concepts that could be taught repeatedly through hands-on, basic application. That workshop was the first of many and I still—ten years later—use the concepts to teach my students. The basic and simple mechanics of science that are incorporated into these lessons reach the gifted and struggling learner and can be adapted and adjusted to time and to the needs of students' individual levels and learning styles."

—Marcie H. Judd, fourth grade teacher, Valley Elementary School,
Kanab School District, Kanab, Utah

"Many elementary teachers lack science background knowledge. The Hands-On Science Activities series provides teachers with both background knowledge and engaging methods to help all students learn science concepts. Marv explains them using simple, clear, and concise language that equips teachers to teach with confidence. These books have increased both the accuracy and the quality of science education."

—Julie Cook, Title I Literacy Coordinator
and former kindergarten teacher,
Logan City School District, Logan, Utah

Jossey-Bass Teacher

Jossey-Bass Teacher provides K–12 teachers with essential knowledge and tools to create a positive and lifelong impact on student learning. Trusted and experienced educational mentors offer practical classroom-tested and theory-based teaching resources for improving teaching practice in a broad range of grade levels and subject areas. From one educator to another, we want to be your first source to make every day your best day in teaching. *Jossey-Bass Teacher* resources serve two types of informational needs—essential knowledge and essential tools.

Essential knowledge resources provide the foundation, strategies, and methods from which teachers may design curriculum and instruction to challenge and excite their students. Connecting theory to practice, essential knowledge books rely on a solid research base and time-tested methods, offering the best ideas and guidance from many of the most experienced and well-respected experts in the field.

Essential tools save teachers time and effort by offering proven, ready-to-use materials for in-class use. Our publications include activities, assessments, exercises, instruments, games, ready reference, and more. They enhance an entire course of study, a weekly lesson, or a daily plan. These essential tools provide insightful, practical, and comprehensive materials on topics that matter most to K–12 teachers.

Hands-On Physical Science Activities for Grades K–6

Second Edition

Marvin N. Tolman, Ed.D.

JOSSEY-BASS
A Wiley Imprint
www.josseybass.com

Published by Jossey-Bass
A Wiley Imprint
989 Market Street, San Francisco, CA 94103-1741   www.josseybass.com

Jossey-Bass books and products are available through most bookstores. To contact Jossey-Bass directly call our Customer Care Department within the U.S. at 800-956-7739, outside the U.S. at 317-572-3986, or fax 317-572-4002.

Jossey-Bass also publishes its books in a variety of electronic formats. Some content that appears in print may not be available in electronic books.

ISBN-13: 978-0-7879-7867-9
ISBN-10: 0-7879-7867-1

*I dedicate this book to
the late Dr. James O. Morton, my mentor, dear friend, and
coauthor of* The Science Curriculum Activities Library, *from which the current series evolved.*

Acknowledgments

Mentioning the names of all individuals who contributed to *The Science Problem-Solving Curriculum Library* would require an additional volume. The author is greatly indebted to the following:

- Teachers and students of all levels.
- School districts throughout the United States who cooperated by supporting and evaluating ideas and methods used in this book.
- The late Dr. James O. Morton, my mentor, dear friend, and co-author of *The Science Curriculum Activities Library*, from which the current series evolved.
- Dr. Garry R. Hardy, my teaching partner for many years, for his constant encouragement and creative ideas.
- Finally, my angel Judy, for without her love, support, encouragement, patience, and acceptance, these books could never have been completed.

Contents

Section One: The Nature of Matter

Section Two: Energy

Section Three: Light

Section Four: Sound

Section Five: Simple Machines

Section Six: Magnetism

Section Seven: Static Electricity

Section Eight: Current Electricity

About the Author

Dr. Marvin N. Tolman trained as an educator at Utah State University and began his career as a teaching principal in rural southeastern Utah. The next eleven years were spent teaching grades one through six in schools of San Juan and Utah Counties and earning graduate degrees.

Currently professor of elementary education, Dr. Tolman has been teaching graduate and undergraduate science methods courses at Brigham Young University since 1975. He has served as a consultant to school districts, taught workshops in many parts of the United States, and published numerous articles in professional journals. With Dr. James O. Morton, Dr. Tolman wrote the three-book series of elementary science activities called *The Science Curriculum Activities Library,* published in 1986.

Dr. Tolman now lives with his wife, Judy, in Spanish Fork, Utah, where they have raised five children.

About the Library

The Science Problem-Solving Curriculum Library evolved from an earlier series by the same author, with Dr. James O. Morton as coauthor: *The Science Curriculum Activities Library*. The majority of the activities herein were also in the earlier publication, and the successful activity format has been retained. Many activities have been revised, and several new activities have been added. A significant feature of this series, which was not in the original, is a section called "For Problem Solvers," included in most of the activities. This section provides ideas for further investigation and related activities for students who are motivated to extend their study beyond the procedural steps of a given activity. We hope that most students will pursue at least some of these extensions and benefit from them.

The Science Problem-Solving Curriculum Library provides teachers with hundreds of science activities that give students hands-on experience related to many science topics. To be used in conjunction with whatever texts and references you have, the Library includes three books, each providing activities that explore a different field. The books are individually titled as follows:

- *Hands-On Life Science Activities for Grades K–6*
- *Hands-On Physical Science Activities for Grades K–6*
- *Hands-On Earth Science Activities for Grades K–6*

More than ever before, children of today grow up in a world impacted by science and technology. A basic understanding of nature and an appreciation for the world around them are gifts too valuable to deny these precious young people, who will be the problem solvers of tomorrow. In addition, a strong science program with a discovery/inquiry approach can enrich the development of mathematics, reading, social studies, and other areas of the curriculum. The activities in the Library develop these skills. Most activities call for thoughtful responses, with questions that encourage analyzing, synthesizing, and inferring, instead of simply answering yes or no.

Development of thinking and reasoning skills, in addition to learning basic content information, are the main goals of the activities outlined herein. Learning how to learn, and how to apply the various tools of learning, are more useful in a person's life than is the acquisition of large numbers of scientific facts. Students are encouraged to explore, invent, and create as they develop skills with the processes of science. The learning of scientific facts is a by-product of this effort, and increased insight and retention associated with facts learned are virtually assured.

How to Use This Book

This book consists of more than 175 easy-to-use, hands-on activities in the following areas of earth sciences:

- Nature of Matter
- Energy
- Light
- Sound

- Simple Machines
- Magnetism
- Static Electricity
- Current Electricity

Use in the Classroom

The activities in this book are designed as discovery activities that students can usually perform quite independently. The teacher is encouraged to provide students (usually in small groups) with the materials listed and a copy of the activity from the beginning through the "Procedure." The section titled "Teacher Information" is not intended for student use, but rather to assist the teacher with discussion following the hands-on activity, as students share their observations. (Be sure to cover this section before copying.) Discussion of conceptual information prior to completing the hands-on activity can interfere with the discovery process.

Correlation with National Standards

The National Research Council produced the *National Science Education Standards*. Published in 1996 by the National Academy of Sciences, this document has become the standard by which many sets of state and local science standards have been developed since that time. The document is available from the National Academy Press, 2101 Constitution Avenue, NW, Box 285, Washington, DC 10055. It is also available online by searching for National Science Education Standards.

For grades K–12 there are seven Standards, identified as Standards A–G. Within each grade-level cluster (K–4, 5–8, 9–12) general subparts are identified.

Content Standard A: Science as Inquiry
Content Standard B: Physical Science
Content Standard C: Life Science
Content Standard D: Earth and Space Science
Content Standard E: Science and Technology
Content Standard F: Science in Personal and Social Perspectives
Content Standard G: History and Nature of Science

This book is one of a series of three books that focus separately on the Life, Earth, and Physical Sciences. Following is a list of the Content Standards to which the set of three books relate for Levels K–4 and 5–8 of the National Science Education Standards. Collectively, the Standards identified for correlation purposes are Standards A, B, C, D, and F. These Standards and their subparts are listed below. In the Standards document, the subparts are in a bulleted list. They are numbered here to facilitate referencing them individually within this book. The appropriate Standards and subparts are repeated in each section of all three books to provide a section-by-section correlation with the National Standards. This information is found on the page immediately preceding the first activity of each section in each book. Since the activities are designed for inquiry, you will note that Content Standard A is included in all sections of each book.

K–4: Content Standard A: Science as Inquiry
As a result of activities in grades K–4, all students should develop:

1. Abilities necessary to do scientific inquiry
2. Understanding about scientific inquiry

K–4 Content Standard B: Physical Science
As a result of activities in grades K–4, all students should develop understanding of

1. Properties of objects and materials
2. Position and motion of objects
3. Light, heat, electricity, and magnetism

K-4 Content Standard C: Life Science

As a result of activities in grades K-4, all students should develop understanding of

1. The characteristics of organisms
2. Life cycles of organisms
3. Organisms and environments

K-4 Content Standard D: Earth and Space Science

As a result of activities in grades K-4, all students should develop understanding of

1. Properties of earth materials
2. Objects in the sky
3. Changes in earth and sky

K-4 Content Standard F: Science in Personal and Social Perspectives

As a result of activities in grades K-4, all students should develop understanding of

1. Personal health
2. Characteristics and changes in populations
3. Types of resources
4. Changes in environments
5. Science and technology in local challenges

5-8 Content Standard A: Science as Inquiry

As a result of activities in grades 5-8, all students should develop:

1. Abilities necessary to do scientific inquiry
2. Understanding about scientific inquiry

5-8 Content Standard B: Physical Science

As a result of activities in grades 5-8, all students should develop understanding of

1. Properties and changes of properties in matter
2. Motions and forces
3. Transfer of energy

5-8 Content Standard C: Life Science

As a result of activities in grades 5-8, all students should develop understanding of

1. Structures and function in living systems
2. Reproduction and heredity
3. Regulation and behavior
4. Populations and ecosystems
5. Diversity and adaptations of organisms

5-8 Content Standard D: Earth and Space Science

As a result of activities in grades 5-8, all students should develop understanding of

1. Structure of the earth system
2. Earth's history
3. Earth in the solar system

5-8 Content Standard F: Science in Personal and Social Perspectives

As a result of activities in grades 5-8, all students should develop understanding of

1. Personal health
2. Populations, resources, and environments
3. Natural hazards
4. Risks and benefits
5. Science and technology in society

Teacher Qualifications

Two important qualities of the elementary teacher as a scientist are (1) commitment to helping students acquire learning skills and (2) recognition of the value of science and its implications in the life and learning of the child.

You do not need to be a scientist to conduct an effective and exciting science program at the elementary level. Interest, creativity, enthusiasm, and willingness to become involved and try something new are the qualifications the teacher of elementary science needs most. Your expertise will grow and your "comfort zone" will expand as you teach, and as you find opportunities to expand your knowledge in the topics you teach. If you haven't yet really tried teaching hands-on science, you will find it to be a lot like eating peanuts—you can't eat just one. Try it. The excitement and enthusiasm you see in your students will bring you back to it again and again.

Early Grades

Many of these concrete activities are easily adaptable for children in the early grades. Although the activity instructions ("Procedures") are written for the student who can read and follow the steps, that does not preclude teachers of the lower grades from using the activities with their students. With oral instructions and slight modifications, many of these activities can be used with kindergarten, first-grade, and second-grade students. In some activities, steps that involve procedures that go beyond the level of the child can simply be omitted and yet offer the child an experience that plants the seed for a concept that will germinate and grow later on.

Teachers of the early grades will probably choose to bypass many of the "For Problem Solvers" sections. That's okay. The "For Problem Solvers" sections are provided for those who are especially motivated and want to go beyond. Use the basic activities for which procedural steps are written and enjoy worthwhile learning experiences together with your young students.

Capitalize on Interest

These materials are both nongraded and nonsequential. Areas of greatest interest and need can be emphasized. As you gain experience with using the activities, your skill in guiding students toward appropriate discoveries and insights will increase.

Organizing for an Activity-Centered Approach

Current trends encourage teachers to use an activity-based program, supplemented by the use of textbooks and many other reference materials, including those available on the Internet. The activities herein encourage hands-on discovery, which enhances the development of valuable learning skills through direct experience. Opportunities abound for students to work together, and such collaboration is encouraged throughout.

One of the advantages of this approach is the elimination of the need for all students to have the same book at the same time, freeing a substantial portion of the textbook money for purchasing a variety of materials and references, including other textbooks, trade books, audio- and videotapes, videodiscs, models, and other visuals. References should be acquired that lend themselves developmentally to a variety of approaches, subject-matter emphases, and levels of reading difficulty.

Grabbers

The sequence of activities within the sections of this book is flexible and may be adjusted according to interest, availability of materials, time of year, or other factors. Many of the activities in each section can be used independently as *grabbers,* to capture student interest. Used this way, they can help to achieve several specific objectives:

- To assist in identifying student interests and selecting topics for study.
- To provide a wide variety of interesting and exciting hands-on activities from many areas of science. As students investigate activities that are of particular interest, they will likely be motivated to try additional related activities in the same section of the book.
- To introduce teachers and students to the discovery/inquiry approach.
- To be used for those occasions when only a short period of time is available and a high-interest independent activity is needed.

Unique Features

The following points should be kept in mind while using this book:

1. Most of these activities can be used with several grade levels, with little adaptation.
2. The student is the central figure when using the discovery/inquiry approach to hands-on learning.
3. The main goals are problem solving and the development of critical-thinking skills. The learning of content is a spinoff, but it is possibly learned with greater insight and meaning than if it were the main objective.

4. It attempts to prepare teachers for inquiry-based instruction and to sharpen their guidance and questioning techniques.

5. Most materials needed for the activities are readily available in the school or at home.

6. Activities are intended to be open and flexible and to encourage the extension of skills through the use of as many outside resources as possible: (a) The use of parents, aides, and resource people of all kinds is recommended throughout; (b) the library, media center, and other school resources, as well as classroom reading centers related to the areas of study, are essential in the effective teaching and learning of science; and (c) educational television, videos, and Internet resources can greatly enrich the science program.

7. With the exception of the activities labeled "teacher demonstration" or "whole-class activity," students are encouraged to work individually, in pairs, or in small groups. In most cases the teacher gathers and organizes the materials, arranges the learning setting, and serves as a resource person. In many instances, the materials listed and the procedural steps are all students will need in order to perform the activities.

8. Information is given in "To the Teacher" at the beginning of each section and in "Teacher Information" at the end of each activity to help you develop your content background and your questioning and guidance skills, in cases for which such help is needed.

9. The last activity of each section, just prior to the "Do You Recall?" questions, is a "Word Search Puzzle," provided especially for those who enjoy word games and building their vocabulary. In addition to providing a word puzzle to be solved, it invites the student to create one of his or her own.

Integrating

Students do more than just science as they become involved in science inquiry activities. Such experiences are loaded with applications of math, reading, and language arts. They often involve social studies, art, music, and physical education as well. This is true of activities throughout this book. Meaningful application and reinforcement of skills learned throughout the curriculum are embedded in the child's science experiences. These connections are noted within the individual activities, as described under "Format of Activities."

Format of Activities

Each activity in this book includes the following information:

- *Activity Number:* Activities are numbered sequentially within each section for easy reference. Each activity has a two-part number to identify the section and the sequence of the activity within the section.

- *Activity Title:* The title of each activity is in the form of a question that can be answered by completing the activity. Each question requires more than a simple yes or no answer.

- *Special Instructions:* Some activities are intended to be used as teacher demonstrations or whole-group activities, or they require close supervision for safety reasons, so these special instructions are noted.

- *Take Home and Do with Family and Friends:* Many activities could be used by the student at home, providing enjoyment and learning for others in the family. Such experiences can work wonders in the life of the child, as he or she teaches others what has been learned at school. The result is often a greater depth of learning on the part of the child, as well as a boost to his or her self-esteem and self-confidence. An activity is marked as "Take home and do with family and friends" if it meets all of the following criteria:

  (1) It uses only materials that are common around the home.
  (2) It has a high chance of arousing interest on the part of the child.
  (3) It is safe for a child to do independently, that is, it uses no flame, hot plate, or very hot water.

  Of course, other activities could be used at the discretion of parents.

- *Materials:* Each activity lists the materials needed. The materials are easily acquired. In some cases special instructions or sources are suggested.

- *Procedure:* The procedural steps are written to the student, in easy-to-understand language. To avoid interfering with the discovery process, no conceptual information is given in this section. This section is not intended for student use, but rather to assist the teacher with discussion following the hands-on activity, as students share their observations. Discussion of conceptual information prior to completing the hands-on activity can interfere with the discovery process.

- *For Problem Solvers:* Most activities include this section, which suggests additional investigations or activities for students who are motivated to extend their study beyond the activity specified in the procedural steps.
- *Teacher Information:* Suggested teaching tips and background information are given. This information supplements that provided in "To the Teacher" at the beginning of each section.
- *Integrating:* Other curricular areas from which students will likely use skills as they complete the activity are recognized here.
- *Science Process Skills:* This is a list of the science process skills likely involved in doing the activity.

Use of Metric Measures

Most measures used are given in metric units followed by units in the English system in parentheses. This is done to encourage use of the metric system.

Word Search Puzzles

The final activity of each section is a "Word Search," intended for fun and for encouraging vocabulary development. Each Word Search includes an answer key and a challenge for students to create word search puzzles of their own, with a form provided for ease of use.

Assessment and Evaluation

The activities in this book should effectively supplement and enrich your science curriculum. While a comprehensive testing system is not provided, a small collection of questions, titled "Do You Recall?" is included at the end of each section. The "Do You Recall" questions could be used as short quizzes or added to other assessments such as pretests or posttests. These may include true/false, multiple-choice, and short-answer questions. No performance assessment items are included. Each list of questions is followed by an answer key, which identifies the related learning activities as well as provides answers to the questions.

Grade Level

The activities in this book are intended to be nongraded. Many activities in each section can be easily adapted for use with young children, while other activities provide challenge for the more talented in the intermediate grades.

Scientific Investigation

Scientists sometimes seem to flip-flop with theories that support a particular health remedy or behavior, then refute it, or alter their position on an established scientific theory. If we understand why that happens, we are less likely to be annoyed with it. It isn't because they are an indecisive lot, but because they are doing what scientists do. If we know of a certain drug that shows promise of helping us with a serious health problem, but is undergoing extended testing prior to approval for public use, we become impatient with the system. If it is marketed without sufficient proof of safety, and we suffer ill side effects, we bring a class-action lawsuit against those who marketed it with insufficient testing.

The Tentative Nature of Scientific Knowledge

Scientific knowledge is, by its very nature, tentative and subject to scrutiny and change. It results from empirical evidence, but the interpretation of that evidence is often unavoidably influenced by human observation, inference, and judgment. To be respected by the scientific community, evidence must usually be verified by multiple scientists.

Occasionally, a long-held theory is shown to be flawed and, in light of new information, must be altered or abandoned. In 1633 Galileo was punished and imprisoned for defying the commonly held "truth" that the earth is the center of the universe and that all celestial bodies revolve around it.

If we criticize change that is based on the best current information, we make light of a process that deserves, instead, to be applauded, and we show our ignorance regarding the scientific way of learning. If we understand and encourage the continuous cyclical process of acquiring new information, we will not be frustrated when later on the "new" theory is also adjusted in light of newer data and revised perspectives.

We are bombarded with so-called "scientifically proven" health remedies that are at odds with each other. Two conflicting theories can both be wrong, but they cannot both be right. In such cases, we must wonder why more data were not collected before certain claims were made, and we question the motive that inspired the marketing of the product.

Taking a Less Formal Approach

In our efforts to spawn lifelong learners, we need to capture the general concept that science is an ongoing search for observable and reproducible evidence. Current thinking is leading us away from the once-held rigid sequence of memorized steps of the scientific method. We are, instead, encouraged to think of scientific investigation as a logical process of acquiring information and solving problems, applicable and useful in virtually every area of study and every walk of life.

In the face of a dilemma brought about by a question or problem, we need to first examine the information currently available. If this is insufficient, we pursue possible ways to acquire additional information (data) and to determine the accuracy of the new data. Having done that, we are better informed and thus better prepared to provide an answer to the question or a solution to the problem.

When Mother comes home from shopping and finds milk spilled on the kitchen floor, the problem is immediately apparent. Her mind quickly surveys current information by considering who has probably been in the house and, based on past behavior, age of children, and other information that would indicate who is most likely responsible for the mess. It may be that the child she first suspected is at scout camp and was nowhere around on that day. Mom quickly adjusts her thinking, considers other ideas, and forms a tentative conclusion (hypothesis). As she seeks out the young suspect for clean-up patrol, she might consider alibis, the stories of witnesses, and other new information, all of which contribute to a final conclusion. Her systematic approach might result in evidence that it was Dad, in a hurry to get to work on time!

Although this example does not involve clear-cut scientific experimentation, it does use a systematic procedure for acquiring information and/or arriving at a solution to an everyday problem, which is the very point being made. The procedure is very simple:

1. Examine information currently available regarding the question or problem at hand.
2. Determine the most logical conclusion (hypothesis) from the information already known.
3. Devise a way to test the accuracy of the hypothesis.

The scientific method, in some form, applies to many situations in everyday life, even if it is not commonly recognized as a scientific approach. If students are trained in the use of a logical, simple, systematic way to gather information and solve problems, they will be better equipped to meet life's inevitable dilemmas.

Now let's go back to the drug company that challenges our patience with extended testing delays. The whole process began with a question raised about the possible effectiveness of a particular formula in dealing with a specific health problem or class of problems. Interested scientists likely examined all of the information they could find about the possible effects of the separate ingredients and of this particular mixture of ingredients. They posed their best tentative conclusion (hypothesis), based on the available information. They then devised ways to test the substance for its effectiveness in the intended application, and perhaps other related applications, and the potential side effects. From their new data, they either concluded that their earlier tentative conclusion (hypothesis) was right, or they adjusted it and did more testing and examination of information (data) acquired. They likely will not allow the substance to be marketed as a solution to the targeted health problem until they are quite certain the substance has been adequately proven to be effective and safe—that their hypothesis was right. Even if you and I are impatient with the lengthy process, and the substance shows promise of reducing suffering and saving lives, we won't see it on the market until it has been conclusively proven. Even then, the new formula could later be shown to be inadequate or to have unforeseen harmful side effects, and now it backfires. Those who tested, approved, and marketed the new drug are facing the wrath of those who used it—perhaps many of the same people who criticized them earlier for taking so long in the approval process.

Broad Application

Your students may use this approach in a variety of subject areas. For example, in the science classroom they might apply it as they experiment with the effect of light on plant growth or the effect of vinegar on baking soda and other powders. Perhaps they will find ways to check their hypothesis regarding the direction of the earth's rotation and do numerous other scientific explorations.

In another setting, they might study diacritical markings and their uses in unlocking word pronunciations, making logical conclusions or predictions based on information at hand. Testing such predictions becomes a learning experience as students read and study further, compare known words, ask an expert (perhaps the teacher), and so forth.

Your students might also enjoy the challenge of identifying elements of scientific investigation in a mystery story. When introduced to a variety of experiences in applying a basic approach to problem solving, they will recognize its universal application and down-to-earth logic, and chances are they will use it with confidence throughout their lives.

Application in This Book

Good habits of scientific investigation are encouraged in the following ways:

1. Students are encouraged to work together in completing the activities, collaborating and learning from each other as they go along.

2. The form "Science Investigation Journaling Notes" is provided to help students record what they do and what they observe, as they practice scientific behavior in their study of science. It will serve as a template that can be applied in numerous problem-solving situations.

3. To help students make the transition into the use of journaling notes to record what they do and observe, the "Science Investigation Journaling Notes" form is reproduced within a few activities of the book. In each of these cases, the activity number is referenced in the heading, and part 1, the "Question," is provided on the form. In addition, the procedural steps guide the student in making appropriate connections and filling out the form as they complete the activity. This help is provided with the following activities:

   Section 1 (Nature of Matter): Activities 9, 10

   Section 2 (Energy): Activity 10

   Section 3 (Light): Activities 8, 28

   Section 4 (Sound): Activity 15

   Section 5 (Simple Machines): Activities 1, 4, 5, 10, 12, 17, 20

   Section 6 (Magnetism): Activity 5

4. Help is also provided for making the next step in the transition, designing an investigation and completing the form independently. The following activities suggest, in "For Problem Solvers" and/or at the beginning of the "Procedural Steps," that students obtain a copy of the "Science Investigation Journaling Notes" from the teacher and use it as they design and complete their own science investigation from the ideas provided.

   Section 1 (Nature of Matter): Activities 18, 22, 23, 27

   Section 2 (Energy): Activities 6, 7, 17, 20, 21

   Section 3 (Light): Activities 13, 16, 21, 22, 25

   Section 4 (Sound): Activities 5, 10, 15, 16, 18, 19

   Section 5 (Simple Machines): Activities 2, 3, 4

Section 6 (Magnetism): Activity 15

Section 7 (Static Electricity): Activity 2

Section 8 (Current Electricity): Activity 1

5. We hope that copies of the following "Science Investigation Journaling Notes" will be made available for students to use as needed, and students will prepare Journaling Notes as they complete these and many other science investigations from this book and from other sources, including those devised by the students themselves.

Science Investigation

Journaling Notes

1. Question:

_____

_____

2. What we already know:

_____

_____

3. Hypothesis:

_____

_____

4. Materials needed:

_____

_____

5. Procedure:

_____

_____

6. Observations/New information:

_____

_____

7. Conclusion:

_____

_____

Final Note: Let's Collaborate and Learn Together

Discovering the excitement of science and developing new techniques for critical thinking and problem solving are major goals of elementary science. The activities in this book are written with the intent that students work together in a collaborative way in the learning process. The discovery/inquiry approach, for which these activities are written, also must emphasize verbal responses and discussion. *It is important that students experience many hands-on activities* in the learning of science *and that they talk about what they do.* Working in collaborative groups facilitates this. Each child should have many opportunities to describe observations and to explain what he or she did and why, because it is important to reinforce the connections between student observations and the science concepts involved. With the exception of recording observations, these activities usually do not require extensive writing, but that, too, is a skill that can be enriched through interest and involvement in science.

There is an ancient Chinese saying: "A journey of a thousand miles begins with a single step." May these ideas and activities help to provide that first step—or the appropriate next step—for all who use this book and are traveling the road of lifelong learning and discovery.

Marvin N. Tolman

Key to Icons

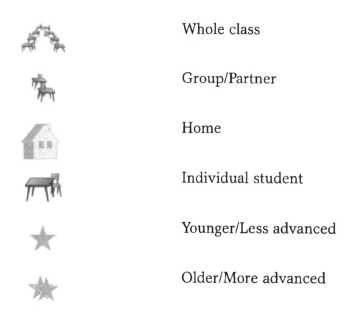

Whole class

Group/Partner

Home

Individual student

Younger/Less advanced

Older/More advanced

Correlation with National Standards Grid

Grades K–4	Section	Pages
A. Science as Inquiry		
1. Abilities necessary to do scientific inquiry	Nature of Matter	1–74
	Energy	75–142
2. Understanding about scientific inquiry	Light	143–216
	Sound	217–280
	Simple Machines	281–365
	Magnetism	367–406
	Static Electricity	407–436
	Current Electricity	437–497
B. Physical Science		
1. Properties of objects and materials	Nature of Matter	1–74
	Energy	75–142
2. Position and motion of objects	Light	143–216
	Sound	217–280
3. Light, heat, electricity, and magnetism	Simple Machines	281–365
	Magnetism	367–406
	Static Electricity	407–436
	Current Electricity	437–497

C. Life Science
    1. Characteristics of organisms
    2. Life cycles of organisms
    3. Organisms and environments

D. Earth and Space Science
    1. Properties of earth materials
    2. Objects in the sky
    3. Changes in earth and sky

E. Science in Personal and Social Perspectives
    1. Personal health
    2. Characteristics and changes in populations
    3. Types of resources
    4. Changes in environments
    5. Science and technology in local challenges

Grades 5–8

A. Science as Inquiry

     1. Abilities necessary to do
         scientific inquiry

     2. Understanding about
         scientific inquiry

Section	Pages
Nature of Matter	1–74
Energy	75–142
Light	143–216
Sound	217–280
Simple Machines	281–365
Magnetism	367–406
Static Electricity	407–436
Current Electricity	437–497

B. Physical Science

     1. Properties of objects and
         materials

     2. Position and motion of objects

     3. Light, heat, electricity,
         and magnetism

Section	Pages
Nature of Matter	1–74
Energy	75–142
Light	143–216
Sound	217–280
Simple Machines	281–365
Magnetism	367–406
Static Electricity	407–436
Current Electricity	437–497

C. Life Science

     1. Characteristics of organisms

     2. Life cycles of organisms

     3. Organisms and environments

D. Earth and Space Science

     1. Properties of earth materials

     2. Objects in the sky

     3. Changes in earth and sky

E. Science in Personal and Social Perspectives

     1. Personal health

     2. Characteristics and changes
         in populations

     3. Types of resources

     4. Changes in environments

     5. Science and technology in
         local challenges

Listing of Activities by Topic

Nature of Matter

Topic	Activities
Acids and Bases	1.25
Atoms and Molecules	1.7, 1.22, 1.24, 1.26, 1.27
Buoyancy	1.9, 1.10
Crystals	1.24
Density	1.9, 1.10
Elements, Compounds	1.13, 1.25
Evaporation and Condensation	1.4, 1.5
Mixtures and Solutions	1.6, 1.7, 1.8, 1.23
Physical and Chemical Changes	1.12, 1.18, 1.19, 1.20, 1.21
Polyethylene	1.17
Solids, Liquids, and Gases	1.3, 1.11, 1.12, 1.14, 1.15, 1.16, 1.17
Surface Tension	1.3
Viscosity	1.28

Energy

Topic	Activities
Center of Gravity	2.21, 2.22, 2.23, 2.24
Centrifugal Force	2.26
Conservation of Energy	2.1, 2.2
Gravity	2.20, 2.21, 2.22, 2.23, 2.24
Heat	2.10, 2.11, 2.12, 2.13, 2.14, 2.17
Inertia	2.25
Kinetic Energy and Potential Energy	2.3, 2.8
Magnetism	2.7
Solar Energy	2.5, 2.15, 2.16, 2.18, 2.19
Sound	2.6
Sources of Energy	2.5, 2.6, 2.7, 2.8, 2.9, 2.10, 2.15
Wind	2.5
Work	2.4

Light

Topic	Activities
Color	3.13, 3.14, 3.15, 3.16, 3.28, 3.29
Lenses	3.23, 3.24, 3.25, 3.26, 3.27
Prisms	3.28, 3.29
Reflection	3.7, 3.8, 3.9, 3.10, 3.11, 3.12
Refraction	3.17, 3.18, 3.19, 3.20, 3.21, 3.22, 3.28
Shadows	3.1, 3.2
Sources	3.6
Transparent/Translucent/Opaque	3.5
Travels in Straight Lines	3.3
Wave Energy	3.4

Sound

Topic	Activities
Amplifying Sound	4.5, 4.14, 4.15, 4.19
Controlling Sound	4.4, 4.20, 4.21
Doppler Effect	4.24
How Sound Travels	4.1, 4.5, 4.6, 4.7, 4.11, 4.13, 4.14, 4.15, 4.17, 4.18
Human Voice	4.8, 4.9, 4.12
Music	
Stringed Instruments	4.4, 4.26
Wind Instruments	4.11, 4.26
Percussion Instruments	4.9, 4.10, 4.24, 4.26
Recognition of Sounds	4.2, 4.3
Sonic Boom	4.25
Speed of Sound	4.16, 4.25
Sympathetic and Forced Vibrations	4.22, 4.23
What Sound Is	4.1, 4.4, 4.5, 4.6, 4.7, 4.14

Machines

Topic	Activities
Friction	5.1, 5.2, 5.3, 5.4
Identifying Machines	5.24
Mechanical Advantage	5.8, 5.9

Simple Machines

Block and Tackle	5.19
First-Class Lever	5.5, 5.6, 5.7, 5.8
Inclined Plane	5.20
Pulley	5.16, 5.17, 5.18, 5.19
Screw	5.22
Second-Class Lever	5.10, 5.11
Third-Class Lever	5.12, 5.13
Wedge	5.21
Wheel and Axle	5.14, 5.15, 5.23

Magnetism

Topic	Activities
Attraction and Repulsion	6.9, 6.10, 6.13
Compasses	6.10, 6.11, 6.12, 6.13
Lodestones	6.1, 6.4
Magnetic Fields	6.13, 6.15, 6.16, 6.17
Magnetic and Nonmagnetic Materials	6.5, 6.6
Making Magnets	6.14
What Magnets Are Like	6.2, 6.3, 6.4, 6.7, 6.8, 6.13

Static Electricity

Topic	Activities
Attraction and Repulsion	7.1, 7.2, 7.3, 7.10, 7.11
Electrostatic Charge	7.2, 7.4, 7.5, 7.6, 7.7, 7.8, 7.9, 7.10, 7.11, 7.13
Induction	7.1, 7.4, 7.5, 7.6, 7.7, 7.8, 7.9, 7.13
Lightning	7.12

Current Electricity

Topic	Activities
Batteries	8.17, 8.18, 8.19
Circuit—What It Is	8.2
Complete and Incomplete Circuits	8.3, 8.5
Conductors and Insulators	8.1
Electricity Can Produce Magnetism	8.9, 8.10, 8.16

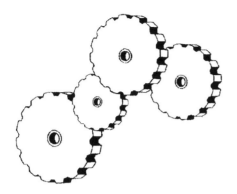

The Nature of Matter

To the Teacher

Everything around us is matter of one form or another. The air we breathe, the food we eat, the books we read, our bodies—all of these things are made of various types of chemicals and substances. The topic of this section is very broad and is related to many other science topics. No attempt has been made to be comprehensive, but only to expose students to a few of the basic properties and relationships of matter. Activities have been selected that involve materials and supplies common to the school or the home, in preference to those requiring sophisticated equipment.

It is recommended that after a study of the nature of matter, you seek opportunities to apply the general concepts learned while studying other science topics. For example, in a study of weather, air, or water, the principles of evaporation and condensation are essential. The effect of temperature change on expansion and contraction is another idea common to weather, air, water, and the topic of this section. In a study of plants, animals, or the human body, the nature of matter has many applications.

The following activities are designed as discovery activities that students can usually perform quite independently. You are encouraged to provide students (usually in small groups) with the materials listed and a copy of the activity from the beginning through the "Procedure." The section titled "Teacher Information" is not intended for student use, but rather to assist you with the discussion following the hands-on activity, as students share their observations. Discussion of conceptual information prior to completing the hands-on activity can interfere with the discovery process.

Regarding the Early Grades

With verbal instructions and slight modifications, many of these activities can be used with kindergarten, first-grade, and second-grade students. With some activities, steps that involve procedures that go beyond the level of the child can simply be omitted and yet offer the child an experience that plants the seed for a concept that will germinate and grow later on.

Teachers of the early grades will probably choose to bypass many of the "For Problem Solvers" sections. That's okay. These sections are provided for those who are especially motivated and want to go beyond the investigation provided by the activity outlined. Use the outlined activities and enjoy worthwhile learning experiences together with your young students. Also consider, however, that many of the "For Problem Solvers" sections can be used appropriately with young children as group activities or as demonstrations, still giving students the advantage of an exposure to the experience, and laying groundwork for connections that will be made later on.

Correlation with National Standards

The following elements of the National Standards are reflected in the activities of this section.

K–4   Content Standard A: Science as Inquiry

As a result of activities in grades K–4, all students should develop:

1. Abilities necessary to do scientific inquiry
2. Understanding about scientific inquiry

K–4   Content Standard B: Physical Science

As a result of activities in grades K–4, all students should develop understanding of

1. Properties of objects and materials
2. Position and motion of objects
3. Light, heat, electricity, and magnetism

5–8   Content Standard A: Science as Inquiry

As a result of activities in grades 5–8, all students should develop:

1. Abilities necessary to do scientific inquiry
2. Understanding about scientific inquiry

5–8   Content Standard B: Physical Science

As a result of activities in grades 5–8, all students should develop understanding of

1. Properties and changes of properties in matter
2. Motions and forces
3. Transfer of energy

How Can You Make a Balance from Two Clothes Hangers?

Materials Needed

- Wire coat hangers
- Skirt hangers with a straight bar and two clips, as shown in Figure 1.1–1.
- Small plastic cups (or paper cups)

Procedure

1. Pull down at the middle of the bottom bar of the coat hanger until the coat hanger is approximately the shape of a rectangle. (See Figure 1.1–2.)

Figure 1.1–1. Skirt Hanger

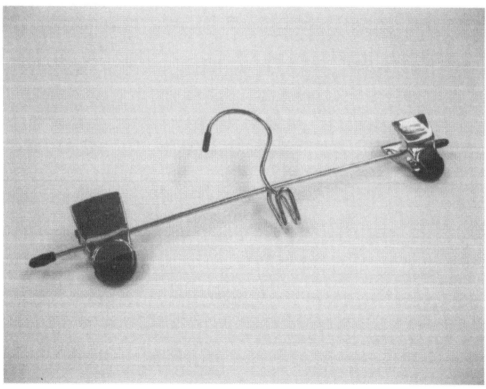

Figure 1.1-2. Preparing the Coat Hanger

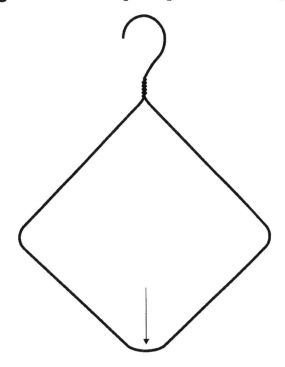

2. Fold the stretched coat hanger forward at the corners so it will stand by itself.
3. Turn the hook around and down, to face inward and upward, as in Figure 1.1-3. This is your wire stand.

Figure 1.1-3. Completed Stand for Two-Hanger Balance

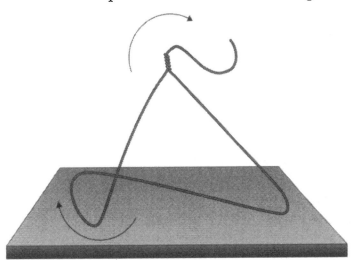

Figure 1.1-4. Complete Two-Hanger Balance

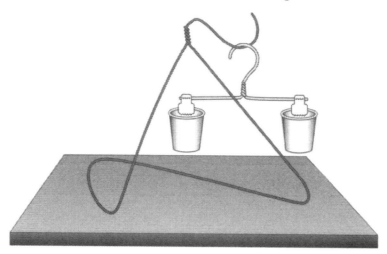

4. Hang the skirt hanger on the wire stand.

5. Fasten one cup to each of the clips of the skirt hanger. This is now your balance beam, and your two-hanger balance is complete and should look like Figure. 1.1-4.

6. Your next task is to balance the balance beam. The clips should both be near the ends of the balance beam (skirt hanger). If necessary, slide them out until they are. Then slide one or both of the clips back and forth slightly until the balance beam is balanced, that is, level.

7. Your two-hanger balance is ready to use.

Teacher Information

Some skirt hangers are designed in such a way that the clips hang lower from the hook. The one that is shown holds the clips higher off the table. Also, most skirt hangers are designed such that the clips slide back and forth to accommodate various sizes of clothing articles. When used as the balance beam for the two-hanger balance, this provides a built-in fine-tuning device for balancing the balance beam before use.

If a skirt hanger is not available, a second wire coat hanger will substitute quite nicely. Simply remove the bottom bar, straighten out the side bars, and install a clothespin on each end such that the clothespin grasps both the balance beam and the cup. Fine-tuning of the balance beam is done by sliding clothespins back and forth as needed, until the two ends are equal distance from the top of the table.

With the cups attached to the clips, and the crossbeam balanced, the two-hanger balance is complete and ready for use. Students will enjoy comparing the masses of small objects. Using nonstandard measures, they can place an object in one cup, then place paper clips or other small objects in the other cup, and then determine how many paper clips or other objects weigh. They will know that an eraser, for instance, weighs a given number of paper clips. If they need to know the actual weight of something, you can provide a set of gram masses. The commercial masses can then be used to make the students' own set of masses by finding an object that balances with each of the needed gram masses and identifying which is 1 gram, which is 2 grams, 5, 10, and so forth.

Integrating
Math, language arts

Science Process Skills
Observing, communicating, measuring

How Can You Make a More Precise Balance from a Ruler?

(Teacher-directed activity)

Materials Needed

- Copies of Figure 1.2–1 for each small group
- Wooden rulers (fairly sturdy)
- Small plastic cups (or paper cups)
- Toothpicks (wooden)
- String
- Paper punches
- Latex cement
- Drill with 1/8-inch bit

Procedure

1. Construct the ruler balance as shown in Figure 1.2-1.

2. Punch three holes near the top of each cup and equal distance apart.

3. Drill two holes near the bottom edge of each end of the ruler. Each pair of holes should be about 2 cm (3/4 in.) apart. The cups will be suspended from these positions.

4. Drill one hole near the top edge of the ruler exactly in the middle of the ruler. The ruler will be suspended from this position.

5. Tie a piece of string (about 30 cm [12 in.]) to the hole at the middle of the ruler.

6. Tie a piece of string (about 20 cm [8 in.]) to one end of the ruler. Loop the string around the two holes in the ruler and tie it quite close to the ruler.

7. Do the same at the other end of the ruler with another piece of string.

8. Cut three pieces of string about 20 cm (8 in.) long. Make a cup hanger by tying one end of each string to one of the three holes in one cup. Tie the other end of these three strings together.

9. Do the same with the other cup.

10. Tie the string that is attached to one end of the ruler to the knot in the three strings of one cup. Tie the other cup to the other end of the ruler in the same way.

11. Hang the ruler balance by tying the middle string to an upper support, such as the handle of a cabinet door.

12. The ruler is your balance beam. Balance it by sliding one or both of the end strings back and forth slightly between the two holes they are attached to.

13. When the balance beam is level, glue the toothpick at the middle of the ruler pointing upward. The toothpick should be in line with the support string. Let the glue dry thoroughly before you use the ruler balance or move it.

14. Your ruler balance is complete and ready to use. Each time you use it, hang it from an upper support and set the balance beam in balance by adjusting the end strings until the toothpick is pointing straight up, in line with the support string.

Teacher Information

This is another way to construct a simple balance scale. This one is a little more difficult to construct; it is also more precise. As with the two-hanger balance, students will enjoy using nonstandard measures to compare masses (weights) of different objects and making their own sets of standard masses. They will know that an eraser, for instance, weighs a given number of paper clips, marbles, pennies, or other item. If they need to know the actual weight of something, you can provide a set of gram masses. The commercial masses can then be used to make the students' own set of masses by finding an object that balances with each of the needed gram masses and identifying which is 1 gram, which is 2 grams, 5, 10, and so forth.

Integrating
Math, language arts

Science Process Skills
Observing, inferring, comparing and contrasting

Figure 1.2-1. Ruler Balance

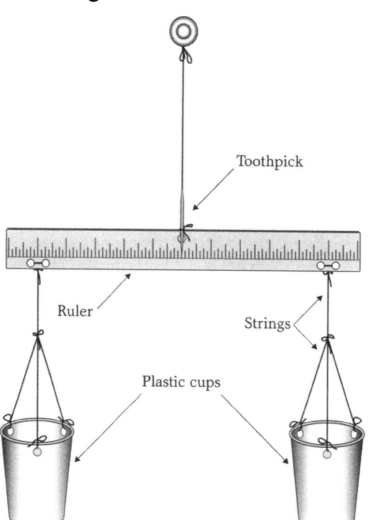

Toothpick

Ruler

Strings

Plastic cups

What Is the Shape of a Drop of Water?

(Take home and do with family and friends.)

Materials Needed

- Waxed paper
- Water
- Eyedroppers
- Pencils

Procedure

1. Draw some water into the dropper.
2. Put several drops on the waxed paper, keeping each separate from the others. Hold the dropper about 1 cm (1/2 in.) above the waxed paper as you squeeze lightly on the bulb.
3. Examine the drops of water. What is their shape? How do they compare in size?
4. Put the point of your pencil into a drop of water, observing carefully to see how the water responds. What did the drop of water do at the surface? Do the water molecules seem to be more attracted to the pencil lead or to each other?
5. Push one of the drops around with your pencil point, observing its behavior.
6. Push two drops together, then three or four. What did they do as they came near each other?
7. What can you say about the attraction of water molecules for each other? For the pencil lead? For the waxed paper?
8. If you could put a drop of water out in space where there is no gravity, what do you think it would look like? What shape would it have? Discuss your ideas with your group.

For Problem Solvers

Place a drop of water on various surfaces and examine each drop with a hand lens. Try it on a sheet of plastic, a sheet of paper, and aluminum foil. Try it on a paper towel. What is the shape of the drop of water? What differences do you see?

Place a drop of water on a penny. Now what is its shape? How many more drops do you think you could put on the penny? Write your prediction, then try it and test your prediction.

Teacher Information

Water molecules attract one another. The attraction of like molecules for one another is called cohesion. Within the liquid, the force of this attraction is balanced, as each molecule is attracted by other molecules all the way around. The molecules on the surface are pulled downward and sideways by neighboring water molecules. This creates a skin-like effect on the surface, called "surface tension." The roundness of the drops on the waxed paper is an indication of surface tension. When several drops are put together they flatten out more, because of the increased effect of gravity as the drop gets larger.

Because of surface tension, a drop of water in the absence of gravity would take on the shape of a perfect sphere. However, a drop of water out in space would evaporate almost instantly.

The surface tension of water forms a bond strong enough to support the weight of a paper clip laid carefully on the water. When a drop of detergent is added to the water, the surface tension is broken and the floating object will sink.

Many other substances have surface tension. It has been suggested that better ball bearings could be formed in space than in factories on the earth because a drop of molten steel would naturally form a perfect sphere.

Integrating

Math, language arts

Science Process Skills

Observing, inferring, predicting, communicating

How Dry Can You Wring a Wet Sponge?

(Take home and do with family and friends.)

Materials Needed

- Meter stick (or yardstick)
- Sponge
- Paper and pencil
- String
- Water

Procedure

1. Wet the sponge, then wring all the water you can out of it.
2. Tie the sponge to one end of the meter stick.
3. Tie a string near the middle of the meter stick; then suspend it by tying it to something overhead. Slide the string on the meter stick to cause it to hang level, as in Figure 1.4–1.
4. Record the time and draw a picture of the setup as it appeared when you prepared it.

Figure 1.4-1. Balanced Meter Stick

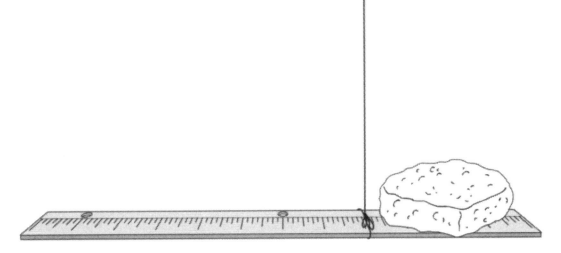

The Nature of Matter

5. Every 15 minutes for two hours, record the time and draw a picture of the setup.

6. What happened to the meter stick during the two hours? Explain why you think this happened. How can you find out whether you were right?

For Problem Solvers

How much water is in a sponge after you wring it out as dry as you can? Can you measure it? *Clue:* How much does the sponge weigh?

After you figure that out, find out how much water is in your bath towel after you dry yourself following a shower. If you live in a humid area, find out whether there is more water in a "dry" towel within a few hours after the towel comes out of the dryer than there was right after the towel was dried. Think about what you did with the sponge.

Teacher Information

The sponge cannot be wrung completely dry. As it sits for the two-hour period, much if not all, of the remaining moisture will evaporate. As the water evaporates, the sponge becomes lighter and the system will no longer be balanced. The process of drawing the position of the setup several times will make students aware of the change as it is taking place. Students will probably want to feel the sponge. This should be avoided until the end of the investigation because of the risk of sliding the string on the stick and nullifying the results.

Students who do the "For Problem Solvers" section will need access to a gram balance or other sensitive weighing device in order to determine the amount of water that is in the wrung-out sponge. With the bath towel activity, they might be able to detect a before/after difference with their bathroom scales at home. To get an accurate measurement, they will need greater precision—perhaps back to the gram balance. They might be amazed to find out how much water is on their bodies when they get out of the shower or bathtub. To measure the water absorbed from the air by a towel in humid conditions, they will again need the gram balance.

Integrating

Math

Science Process Skills

Observing, comparing and contrasting, measuring, using space-time relationships

What Is Condensation?

(Teacher-supervised activity)

Materials Needed

- Saucepan
- Water
- Pie tin (preferably cold)
- Hot plate

Procedure

1. Put about 1 cm (1/2 in.) of water in the saucepan.
2. Heat the water until it boils.
3. Hold the pie tin over the boiling water.
4. Observe the pie tin carefully. What do you see forming on the bottom of the pie tin? Discuss why you think this happens.

For Problem Solvers

Have you noticed that your bathroom mirror is clean and shiny before you get in the shower or bathtub, but that it gets foggy as the hot water runs? Where does the water come from that gets on the mirror? Try to explain how it happens.

Notice the outside of the windows of cars and houses on a cool summer morning. Often you will find moisture on them. Where did the moisture come from? Investigate and find out how this happened. If you live in a cold climate, you will sometimes find frost on the windows. Where does it come from?

Teacher Information

As water is heated, the rate of evaporation increases. Water molecules acquire additional energy from the heat source, and the speed of the molecules increases. At a certain velocity, the molecules are able to escape the surface of the liquid and go into the air. The pie tin held over the escaping water vapor cools the vapor and causes it to condense into liquid form, as shown by the drops of water forming on the bottom. This process will be speeded up if the pie tin is cooled first.

Similar evidence of condensation can be observed by placing a pitcher (or other container) of ice water out on a table. Water vapor in the nearby air is cooled by the pitcher, and drops of water form on the table's surface.

Integrating

Math, language arts

Science Process Skills

Observing, inferring

What Are Mixtures and Solutions?

Materials Needed

- Glass jars
- Spoons (or stirrers)
- Sugar
- Water
- Marbles or small rocks
- Paper clips
- Toothpicks
- Bits of paper
- Paper and pencils

Procedure

1. Fill two jars about half full of water.
2. Put the marbles, paper clips, toothpicks, and bits of paper in one jar and a spoonful of sugar in the other jar.
3. Stir both jars and observe what happens to the materials in the water.
4. Compare the results in the two jars. One is a mixture and the other is a solution.
5. Try other substances in water, such as sand, powdered milk, or powdered chocolate. Make a list of those you think will produce mixtures and those that will produce solutions. Explain the differences you observe.

For Problem Solvers

Try to identify mixtures and solutions that are already in your environment. What about the soil in a flowerbed at home or at school? What about the air you breathe? Make a list of all the mixtures you can find in nature and another list of all the solutions you can find. Notice different food products in your cupboards at home or on grocery-store shelves. Add these to your lists.

Teacher Information

A mixture consists of two or more substances that retain their separate identities when mixed together. Solutions result when the substance placed in a liquid seems to become part of the liquid. A solution is really a special kind of mixture—one in which the particles are all molecular in size.

Materials listed can easily be substituted or supplemented with other soluble and nonsoluble materials.

Integrating

Math, language arts

Science Process Skills

Observing, inferring, classifying, comparing and contrasting

How Can You Separate a Mixture of Salt and Pepper?

(Take home and do with family and friends.)

Materials Needed

- Plastic bag
- One-half cup of salt
- 1 teaspoon of pepper

Procedure

1. Mix the salt and the pepper together in the bag.
2. Now your challenge is to get the pepper out of the salt. How might you do it?
3. Test your ideas, and discuss your observations with your group.

Teacher Information

This activity is intended to help children learn to conduct and evaluate problem-solving procedures. There is no "right" answer but some procedures may be more effective or efficient than others. For example, picking the pepper out is slow and tedious, but it works. Dissolving the salt in water and straining the solution through a cloth is more efficient. Encourage the children to think of and try as many ways as possible. This could introduce a discussion of the way science and technology have combined to find easier and more efficient ways to do things.

Integrating

Math, language arts

Science Process Skills

Observing, inferring

What Happens to Water When You Add Salt?

Materials Needed

- One hard-boiled egg for each small group
- One raw egg for each small group
- Large cups
- Salt
- Measuring spoons

Procedure

1. One of the eggs is raw and the other hard-boiled. Can you tell which of the eggs you received is the hard-boiled egg?
2. Put both eggs in the cup and fill the cup with water.
3. Add salt to the cup, a tablespoon at a time, stirring until the salt dissolves, until something happens to the eggs.
4. What happened?
5. Why do you think salt makes this difference? Share your ideas with other students, and learn from one another.

For Problem Solvers

Did the two eggs respond to the salt water at the same time, or did one of them require more salt than the other? If they were different, which one responded first? Try it again and again, if necessary, until you are sure.

Try spinning a raw egg and a boiled egg. Do they spin equally well? If not, which one spins better? Why do you think it does?

Teacher Information

When the eggs are put in the untreated water, they will both sink to the bottom of the cup. As salt is added to the water, the eggs will rise to the top. This is because salt increases the density of the water. The eggs are more dense than tap water, but less dense than salt water.

Floating an egg in brine solution is the method some people use to tell when the brine is just right for pickling.

It is hoped that your problem solvers noticed that the raw egg rises before the boiled egg does. For these students, the activity shows not only that salt water is more dense, and therefore more buoyant, but that boiled eggs are more dense than raw eggs. The challenge for these young researchers is to find out why.

Hard-boiled eggs also spin better than raw ones. Have students spin a raw egg, stop it, and let it go, and it will begin to spin again. Because of inertia, the inside of the egg continues to move after the outside of the egg stops. Sometimes you can hear or feel the inside of a raw egg if you shake it. The boiled egg is solid, so the spinning action isn't affected by movement of materials inside the egg.

Integrating
Math, reading, language arts

Science Process Skills
Observing, inferring, measuring, communicating, comparing and contrasting, researching

How Does a Hydrometer Work?

(Take home and do with family and friends.)

Materials Needed

- A copy of the "Science Investigation Journaling Notes" for this activity for each student
- Lipstick tube cap (or small test tube)
- Several BBs (or other small weights)
- Tape (or gummed label)
- Plastic tumbler
- Water
- Salt
- Variety of liquids
- Marker
- Paper and pencil

Procedure

1. As you complete this activity you will keep a record of what you do, just as scientists do. Obtain a copy of the form "Science Investigation Journaling Notes" from your teacher and write the information that is called for, including your name and the date.

2. For this activity you will learn what a hydrometer is and how it works. For item 1, the question is provided for you on the form.

3. Item 2 asks for what you already know about the question. If you know anything about hydrometers, write your ideas.

4. For item 3, write a statement of how hydrometers work, or how you think they might work, according to what you already know, and that will be your hypothesis.

5. Now continue with the following instructions. Complete your Journaling Notes as you go. Steps 6 through 12 below will help you with the information you need to write on the form for items 4 and 5.

6. Fill the tumbler about two-thirds full of water.

7. Place a few BBs in the lipstick cap.

8. Put the gummed label or a piece of tape lengthwise on the lipstick cap.

9. Place the lipstick cap, open end up, in the glass of water. Add or remove weights until the cap floats vertically, as shown in Figure 1.9–1.

10. Mark the water level on the cap.

11. Your cap can now be used as a hydrometer. Hydrometers are used for measuring density of liquids, comparing them with the density of water. If the density of another liquid is greater than that of water, the cap will float higher than it floats in water. If the density of the other liquid is less than that of water, the cap will sink deeper into the water.

12. Dissolve about one-fourth cup of salt in your glass of water. Without changing the number of weights in your hydrometer, put the hydrometer in the salt water. Does your hydrometer float deeper than it did in plain water, or does it float higher? What does this tell you about the density of salt water?

13. Complete your "Science Investigation Journaling Notes." Are you ready to explain what a hydrometer is and how it works? Discuss it with your group.

Figure 1.9-1. A Hydrometer.

For Problem Solvers

Obtain several plastic tumblers and fill them about half full, each with a different liquid, such as water, salt water, cooking oil, and rubbing alcohol. Line up the containers in order of least dense to most dense, according to your prediction. Use your hydrometer to compare the liquids, and then arrange the liquids in order, with the liquid of least density on the left and the liquid of greatest density on the right. Did you predict them correctly?

Gather a variety of small objects that are made of plastic, wood, or metal. Place these items in the liquids, one at a time. Try to find at least one item that will float on each liquid but not on the next liquid to the left.

Ask a mechanic (or anyone who services his or her own car) to show you the instrument they use to test radiator fluid. What is it called? Think about it in terms of this activity. Compare it to your hydrometer and explain similarities and differences.

Teacher Information

Any object that floats displaces an amount of liquid equal to its own weight (Archimedes' Principle). If the specific gravity (density) of the liquid is less than that of water, the object floats deeper into the surface of the liquid, as the object has to displace more liquid to equal its own weight. If the hydrometer floats higher, it is in a liquid of greater density.

Hydrometers are used to test such liquids as antifreeze and battery acid, following the Archimedes' Principle.

Integrating

Math, language arts

Science Process Skills

Observing, inferring, classifying, measuring, predicting, communicating, comparing and contrasting

Science Investigation

Journaling Notes for Activity 1.9

1. Question: *How does a hydrometer work?*

2. What we already know:

_____

_____

3. Hypothesis:

_____

_____

4. Materials needed:

_____

_____

5. Procedure:

_____

_____

_____

6. Observations/New information:

_____

_____

7. Conclusion:

_____

_____

_____

How Can the Depth of a Bathyscaph Be Controlled?

(Teacher-supervised activity)

Materials Needed

- Copy of the "Science Investigation Journaling Notes" for this activity for each student
- Fish tank (or bucket) full of water
- Plastic bottle with lid that seals
- Latex cement
- Several marbles (or other weights)
- Plastic tubing
- Drill with a set of bits

Procedure

1. As you complete this activity you will keep a record of what you do, just as scientists do. Use a copy of the form "Science Investigation Journaling Notes" your teacher will give you and write the information that is called for, including your name and the date.

2. For this activity, you will learn how the depth of a bathyscaph can be controlled. For item 1, the question is provided for you on the form.

3. Item 2 asks for what you already know about the question. If you have some ideas about how bathyscaphs work, write your ideas.

4. For item 3, write a statement of how the depth of a bathyscaph can be controlled, according to what you already know, and that will be your hypothesis.

5. Now continue with the following instructions. Complete your Journaling Notes as you go. Steps 6 through 13 below will help you with the information you need to write on the form for items 4 and 5.

6. Drill one hole in the bottle lid. The hole should be just large enough to insert the tube through it.

7. Insert one end of the tubing through the hole in the lid and put latex cement on the lid around the tubing to seal it from leakage of water or air. Allow the cement at least an hour to dry.

8. Drill one tiny hole (about 1/8 inch or smaller) in the bottom of the bottle.

9. Place several marbles in the bottle, put the lid on tight, and place the bottle in the water. The marbles will help to hold the bottom of the bottle down, keeping the tiny hole in the water.

10. Put the end of the tube in your mouth and adjust the amount of water in the bottle by blowing or drawing on the end of the tube until the bottle floats just beneath the surface. The tiny open hole in the bottom of the bottle allows you to control the amount of water in the bottle.

11. Now draw on the tube very slightly to allow a little more water to enter the bottle. What happened to the bottle?

12. Blow and draw on the tube to change the amount of water in the bottle. What happens to the bottle?

13. How could this idea be used in a bathyscaph designed to study the ocean at different depths?

14. Complete your "Science Investigation Journaling Notes." Are you ready to explain how the depth of a bathyscaph can be controlled? Discuss it with your group.

Teacher Information

Caution: Careful supervision is required with the use of the drill.

A vessel in water can be caused to float at different depths by altering the density of the vessel. Density can be increased by displacing a chamber of air (or a portion of it) with water or decreased by displacing the water with air. This is one way the depth of an underwater vessel can be adjusted.

The bottle shown in Figures 1.10–1, 1.10–2, and 1.10–3 has a capacity of approximately 10 ounces. Larger bottles can be used if the tank is large enough to accommodate them. Such bottles are easier to control than are tiny bottles because any given change in the amount of water in the bottle results in a smaller change in the density of the bottle, thus allowing for more control on the "fine-tuning" of the density.

This bit of "technology" cannot be credited to the genius of mankind. It is but one of the many adaptations people have learned from observing nature. Fish are equipped with a swim bladder, a small sac in the abdomen which allows them to maintain buoyancy. The sac inflates if the fish needs to rise in the water and deflates if the fish needs to move to a lower depth.

The hot air balloon also uses the principle of altering density to float at differing altitudes. Instead of adjusting the amount of air to control density, it adjusts the temperature of the air. Temperature change results in density adjustment.

Integrating

Math, language arts, social studies

Science Process Skills

Observing, inferring, predicting, communicating, identifying and controlling variables

Figure 1.10-1. Bathyscaph Lowering into Water

Hands-On Physical Science Activities

Figure 1.10–2.

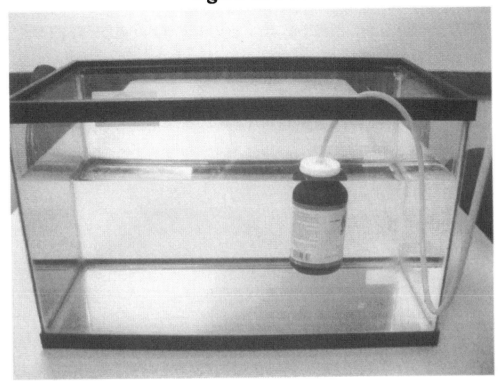

Figure 1.10–3.

Science Investigation
Journaling Notes for Activity 1.10

1. Question: *How can the depth of a bathyscaph be controlled?*

2. What we already know:

_____

_____

3. Hypothesis:

_____

_____

4. Materials needed:

_____

_____

5. Procedure:

_____

_____

_____

6. Observations/New information:

_____

_____

7. Conclusion:

_____

_____

What Are Solids, Liquids, and Gases?

(Teacher-supervised activity)

Materials Needed

- Charcoal briquettes or small pieces of coal
- Hammer
- Safety glasses
- Blocks of wood
- Ice cubes
- Dishes
- Paper towels
- Paper and pencils
- Small plastic bags

Procedure

1. Put the ice cube in the dish and place the block of wood and the dish on a table.
2. Place the paper towel over the block of wood and the charcoal on the paper towel. The purpose of the wood is to provide a pounding block.

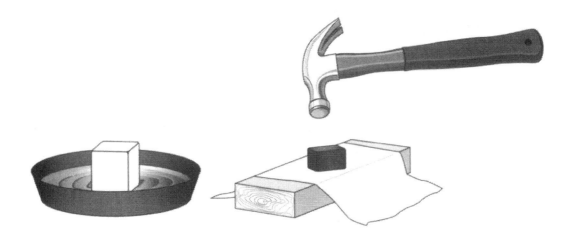

3. Examine and describe the charcoal and the ice cube. How are they alike? How are they different? Tell whether you think each is a solid, liquid, or gas.

4. Put on the safety glasses and crush the piece of charcoal with the hammer. Be sure it is on the paper towel and the wood block as you do this. Pound lightly with the hammer so the pieces don't scatter. For additional safety, you can put the briquette in a plastic bag (such as a sandwich bag) before pounding.

5. Examine the charcoal again. In what ways is it the same as it was before? How is it different? Is it a solid, liquid, or gas?

6. Let the ice cube remain in the dish. Examine and describe it after a few minutes, after an hour, and after a day. Each time, decide whether it is a solid, a liquid, or a gas.

7. After the last observation, again compare the charcoal with the ice cube. How did they respond differently when they were left undisturbed? Why do you think this happened?

8. Make a list of solids, a list of liquids, and a list of gases.

9. Share and discuss your observations with your group.

For Problem Solvers

If ice cubes are placed in room-temperature conditions, they will become liquid before evaporating as a gas. Sometimes, however, ice cubes slowly evaporate as they stand in the ice tray, without even being removed from the freezer. Meat will also eventually dehydrate if left too long in the freezer. See whether you can identify any substance that naturally changes from solid to gas at room temperature, without becoming a liquid.

Do some research and find out how many substances you can identify that exist in nature in all three states—solid, liquid, and gas. How many more can you find that exist in nature as both a solid and a liquid?

Teacher Information

Caution: Safety glasses should be used when crushing the charcoal with the hammer.

All matter is either solid, liquid, or gas. Charcoal remains a solid even when powdered. Water has the unusual property of being easily changed to any of the three states. The solid and liquid states are easily observed.

Point out that when water becomes a gas it is invisible. In many substances, including wood and charcoal, certain elements combine with oxygen when burned and produce a gaseous substance, except for a small amount of ash left behind in solid form. As gases form during burning, visible solid and liquid particles are often suspended in the gases. We call this smoke.

Those who choose to accept the "For Problem Solvers" challenge will truly be challenged. Moth balls change from solid to gas at room temperature without becoming liquid. Perhaps these young scientists will find other substances as well. The only common material that exists in nature in all three states is water.

Integrating

Language arts

Science Process Skills

Observing, predicting, comparing and contrasting, using space-time relationships

How Can You Produce a Gas from a Solid and a Liquid?

(Teacher-supervised activity)

Materials Needed

- Two film canisters (empty) per small group
- Water
- Effervescent pain reliever or antacid tablets
- Paper and pencils

Procedure

1. Place about half of an effervescent tablet in one film canister.
2. Put a small amount of water in the second canister (about one-fourth full).
3. Pour the water from the second canister into the first canister, quickly snap the lid on tight, and stand back!
4. What happened?
5. Did this produce a physical change or a chemical change? How can you tell?
6. Discuss your observations with others in the group. Consider both the type of reaction (physical or chemical) and what you observed in terms of states of matter.

Teacher Information

Consider using this activity as a demonstration, rather than providing printed instructions for students, so you can add the element of surprise. If you are using an antacid tablet, you can easily disguise its appearance by pulverizing it in advance. Be prepared for a bit of excitement. If you snap the lid right back on after it blows, the reaction might recur a time or two.

Have paper towels on hand before you begin. After the reaction, when kids and lids are all gathered up, discuss student observations. Was this a chemical change or a physical change? (The production of a gas is a sure sign of a chemical change.) Also make connections with the concept of states of matter; here we saw a gas produced by combining a solid and a liquid (all three states of matter).

Film canisters vary in color and shape. Try all of the types you can find, and have students keep a record and compare them. Let students help decide what to record and compare. Certainly this should include the type of canister and height of the flight (of the lid, that is, not students). Have them predict the result with each change in variable, and compare actual results with predictions. Predictions should be made by each person (silently predict and write down the prediction) before discussing as a group or class, which will encourage each student to think and reason instead of simply going along with someone else.

You can provide a similar experience with baking soda and vinegar. A third option is dry ice, with which you maintain the thrill of the flying lid and the change of state from a solid to a gas, but you lose both the chemical reaction and the production of a gas from combining a liquid and a solid. *Caution:* Dry ice also involves some safety concerns, such as the possibilities of burned fingers and the temptation to swallow a piece of dry ice.

Integrating

Math, language arts

Science Process Skills

Observing, inferring, comparing and contrasting

How Can You Make a Fire Extinguisher?

(Teacher-supervised activity)

Materials Needed

- Large soda bottles (or quart bottles)
- Vinegar
- Baking soda
- Candles
- Matches
- Sink or pans
- Tablespoons
- Measuring cups

Procedure

1. Stand the candle in the sink or pan. Be sure there are no flammable materials nearby. Light the candle.
2. Put one tablespoon of baking soda into the bottle.
3. Measure about three to four ounces of vinegar with the measuring cup and pour it into the bottle with the baking soda.
4. As bubbles form, hold the bottle over the candle flame and tip it as though you were pouring water from the bottle onto the flame, but do not tip it far enough to pour out the vinegar.
5. What happened to the flame? Explain why you think this happened.

For Problem Solvers

Carbon dioxide is heavier than air. Why does that help it to be effective in putting out fires? See what you can learn about fire fighting and what chemicals are commonly used for putting out fires.

Teacher Information

Caution: This activity must be carefully supervised due to the involvement of fire.

Baking soda is sodium bicarbonate. Vinegar contains acetic acid. When the two mix, carbon dioxide (CO_2) is formed. Carbon dioxide is heavier than air, so when the bottle is tipped, the CO_2 pours out. You don't see it pour because carbon dioxide is colorless. As it pours over the flame, the CO_2 deprives the flame of oxygen and the flame is extinguished. Carbon dioxide is commonly used in some fire extinguishers.

Carbon dioxide is one of the more common gases. Humans and other animals produce it and breathe it into the air. Plants absorb it and, in turn, make oxygen. Carbon dioxide is put into soft drinks to give them bubbles, or fizz. Dry ice is carbon dioxide, frozen to make it solid. If dry ice is available, have students repeat this activity, using a small piece of dry ice in place of vinegar and baking soda. *Caution:* Any use of dry ice must be carefully supervised, as it can burn the skin. It should never be put in the mouth. Also, dry ice must not be placed in a sealed bottle (or any other sealed container), because it can build up enough pressure to explode the container.

Integrating

Reading, math

Science Process Skills

Observing, inferring, measuring, predicting, communicating, identifying and controlling variables, experimenting, researching

How Can a Blown-Out Candle Relight Itself?

(Teacher-supervised activity)

Materials Needed

- Two candles per small group
- Metal pans
- Matches

Procedure

1. For this activity, keep the candles over the pan and be sure your teacher is with you.
2. Light both candles.
3. Hold the two candles horizontally with one flame about an inch above the other, as shown below.

4. Holding both candles steady, blow out the lower flame and observe for a few seconds.
5. What happened? Can you explain why?

For Problem Solvers

Observe the flame of a burning candle very carefully. Where is the flame resting? Does it seem to be sitting right on the wick, and burning the wick, or is it above the wick? What do you think is burning?

See what you can find out about flames. What part of a flame is the hottest? What causes the colors you see in the flame? Find answers to these questions and to other questions you think of.

Teacher Information

Wax, in solid form, does not burn. Heat changes wax to a vapor, which burns when combined with oxygen in the air. When a candle flame is blown out, hot gases continue to rise for a short time. These gases can ignite and act as a wick if another flame is close by and in their path. The flame will burn down the column of gases and relight the lower candle.

Integrating

Reading, language arts, math

Science Process Skills

Observing, inferring, measuring, predicting, communicating, using space-time relationships, formulating hypotheses, identifying and controlling variables, experimenting, researching

How Can You Remove the Flame from a Candle Without Putting It Out?

(Teacher-supervised activity)

Materials Needed

- Glass jars with lid
- Birthday candles
- Tablespoons
- 30 cm (1 ft.) of pliable wire per small group
- Baking soda
- Vinegar
- Matches

Procedure

1. Put two tablespoons of vinegar and one tablespoon of baking soda in the bottom of the jar. Bubbles will form.

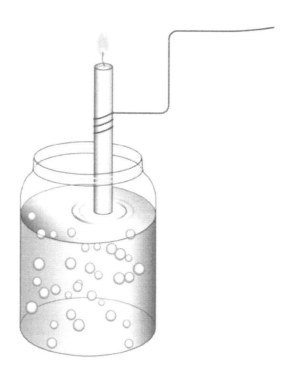

Hands-On Physical Science Activities

2. Set the lid upside down on the jar, to cover the jar without sealing it.

3. Let the jar sit until the bubbling has nearly stopped.

4. While you are waiting for the bubbles to stop, form a holder for the candle from the wire.

5. Place the candle in your wire holder and light the candle.

6. Remove the cover from the jar and slowly lower the candle into the jar until the top of the wick is about an inch below the rim of the jar. Then bring the candle back up.

7. Try it again. Explain what happens.

Teacher Information

Caution: Careful supervision is required due to the involvement of fire. Combining vinegar and baking soda forms carbon dioxide, which is heavier than air and therefore drives the air out of the jar, leaving the jar filled with carbon dioxide. As the flame is lowered below the rim of the jar, it is starved for oxygen and the candle actually burns out. Gases continue to rise from the candle for a short time, however, and the flame sits on top of the layer of carbon dioxide, burning the rising gases in the presence of oxygen.

Integrating

Math, language arts

Science Process Skills

Observing, inferring, measuring, predicting, communicating

How Can You Make a Ball Bounce by Itself?

(Teacher-supervised activity)

Materials Needed

- Old tennis balls
- Scissors

Procedure

1. With the scissors, cut the tennis ball in half to make two dish-shaped halves. Ask your teacher to start the cut by punching a hole with the point of the scissors or with a knife.

2. Trim around the edge of one of the halves until its diameter is about 5 cm (2 in.).

3. Turn the ball dish inside out and set it on the floor or on a table. Observe for several seconds.

4. What happened? Explain why you think it behaved this way. What might you do to make it happen faster or more slowly?

For Problem Solvers

Get your friends to help you ask around for and locate several tennis balls that are old and not needed any more. Experiment with the amount that you trim off for this activity. Try to create the ball that will bounce the highest. See whether you can control the delay time (from the time you set it down until it flips up) by how much of the ball you trim off.

Teacher Information

Caution: You should make the starter holes in the tennis balls to avoid injury to students.

After a brief observation, the "dish" should jump. Rubber molecules act like tiny springs, giving rubber the tendency to spring back to its original shape when distorted. This property gives rubber its bounce. With the inverted "dish," the restoring action of the rubber first has to overcome the resistance of the backward bend. When it reaches a certain point, though, the movement is very quick. The edges strike the surface with considerable force and the ball flips into the air.

As students ponder the last question in step 4, you might need to encourage them to try trimming a little more off the edges of the dish or to take the other half of the ball and trim off less than they did with the first one. Trimming less will delay the action and trimming more will speed it up. Your "problem solvers" are encouraged to investigate with these factors.

Integrating

Math

Science Process Skills

Observing, inferring, classifying, measuring, predicting, communicating, using space-time relationships, formulating hypotheses, identifying and controlling variables, experimenting, researching

What Is Polyethylene?

Materials Needed
- One polyethylene bag with tie per small group
- One nonpolyethylene plastic bag with tie per small group
- Sharpened pencils
- Water
- Sink or large pans

Procedure
1. Check to be sure one of the bags is polyethylene. It will be indicated on the container.
2. Fill both bags with water and put ties around the tops. Keep them over a sink or large pan.
3. Stab the pencil through the nonpolyethylene bag and observe what happens.
4. Stab the pencil through the polyethylene bag. Compare the results with what happened in step 3.
5. Discuss your observations with your group.

Teacher Information
Polyethylene has the strange property of shrinking together when it is torn. When the bag is punctured, the polyethylene shrinks and stops (or reduces) the flow of water. This property is a factor in puncture-resistant tires.

Integrating
Language arts

Science Process Skills
Observing, inferring, classifying, measuring, predicting, communicating, comparing and contrasting, using space-time relationships, formulating hypotheses, identifying and controlling variables, experimenting

Is the Dissolving of Solids a Physical Change or a Chemical Change?

Materials Needed

- Tumblers
- Sugar (or salt)
- Paper and pencils
- Water
- Stirrers
- "Science Investigation Journaling Notes" form for each student

Procedure

1. Do the following as a Science Investigation. Obtain a blank copy of the "Science Investigation Journaling Notes" from your teacher. Write your name, the date, and your question at the top. Plan your investigation through item 5 of the form (Procedure) and have it approved by your teacher. Complete the Journaling Notes as you perform your investigation. When you are finished, share your project with your group, and submit your Journaling Notes to your teacher if requested.

2. Put about two teaspoons of sugar and a small amount of water in a tumbler and stir until the sugar is completely dissolved. Do you think the dissolving of the sugar in the water was a physical change or a chemical change?

3. Put the tumbler where it can remain undisturbed while the water evaporates. Check the tumbler twice each day. Record your observations each time you detect a change.

4. When the water has completely evaporated, record your observations of the tumbler. Do you think the dissolving of the sugar in the water was a physical change or a chemical change? Why do you think as you do? Support your answer with your observations.

For Problem Solvers

Also do this as a Science Investigation. Obtain another copy of the "Science Investigation Journaling Notes." Write your name, the date, and your question at the top. Plan your investigation through item 5 (Procedure) and have it approved by your teacher. Complete the Journaling Notes as you perform your investigation. Share your project with your group, and submit your Journaling Notes to your teacher if requested.

Take a small piece of paper and tear it up into the smallest bits you can. Was that a physical change or a chemical change? Place a drop of lemon juice on a piece of paper and let it dry. Is a physical change or a chemical change taking place? Hold the paper near a light bulb until it begins to look different where the drop of juice was. Is this a physical change or a chemical change? Identify other physical changes and chemical changes that occur in your world. What about a cake as it bakes? What about the soles of your shoes, as they slowly wear away?

Teacher Information

A physical change usually alters only the state of matter, such as from a solid to a liquid or from a liquid to a gas, or the shape, texture, and so on. Physical changes are frequently reversible. For example, water can be obtained by condensing it out of the air or by melting an ice cube. Chemical changes involve changes in molecular structure and are not easily reversible. As the water in this activity evaporates, crystals of sugar appear. They will be massed together and will not look the same, but a taste test will reveal that it is sugar.

You might also burn a bit of sugar for students to compare. After the burned substance has cooled, let someone taste it and determine whether the sugar underwent a physical change or a chemical change. It will no longer taste like sugar, except to the extent that unburned sugar crystals remain. The burning process produces a chemical change. The sugar has been oxidized through heat, leaving a carbon residue.

Integrating

Math

Science Process Skills

Observing, inferring, classifying, measuring, comparing and contrasting

Is Burned Sugar Still Sugar?

(Teacher-supervised activity)

Materials Needed

- Sugar cubes
- Candles
- Matches
- Hand lenses
- Pieces of aluminum foil, about 15 by 15 cm (6 in. by 6 in.)

Procedure

1. Put a sugar cube in the center of one piece of foil.
2. Place the candle in the center of another piece of foil and light the candle with a match.
3. Taking the first piece of foil by the edges, hold it with the sugar cube directly over the lighted candle until some of the sugar turns black.
4. Remove the foil from the flame, set it aside to cool, and blow out the candle.
5. When the burned sugar has cooled, crumble some of it and examine it carefully with the hand lens. Taste some of it.
6. How does the burned sugar compare with sugar that did not burn? Compare both the appearance and the taste of the burned and the unburned sugar.
7. Was this a physical change or a chemical change?
8. Discuss your observations and share your ideas with your group.

Teacher Information

Use caution in having students work with the flame and taste the sugar. This might have to be done as a teacher demonstration in order to comply with the restrictions and guidelines of your school system. Before you ask students to taste anything, be sure there are no food allergies.

To add a measurement experience to this activity, use loose sugar instead of a sugar cube. Have your students measure out one gram of sugar and place the sugar in the center of one piece of foil.

The Nature of Matter

With this activity, most of the hydrogen and oxygen have been driven off into the air as gases. Thus, the burned material left on the foil is carbon, not sugar. Obviously, this is a chemical change. We have no way to recombine oxygen and hydrogen with the carbon and restore the substance as sugar.

As you make connections with science concepts, emphasize that nothing was lost or destroyed, but only changed. The hydrogen, oxygen, and carbon, which together comprised the sugar, still exist. The same atoms that a few minutes earlier formed sugar will likely become a part of many other substances in the future. What a marvelous recycling plan! And incidentally, this recycling process, in one form or another, occurs with almost everything we eat. The body is a chemical plant, and the food we eat is constantly undergoing chemical changes as the body takes the substances we give it and uses them to manufacture the substances we need for life and good health.

Integrating

Math, language arts

Science Process Skills

Observing, inferring, measuring, comparing and contrasting

What Is Rust?

Materials Needed

- Two small identical jars per small group
- Two small identical dishes per small group
- Paper and pencils
- Steel wool
- Water

Procedure

1. Put a small wad of steel wool into one of the jars. Push it clear to the bottom. Pack it just tightly enough that it will stay at the bottom of the jar when the jar is turned upside down.

2. Put about 2 cm (3/4 in.) of water in each of the two dishes. Be sure you put the same amount in each one.

3. Turn the two jars upside down and stand one in each of the dishes. One jar should have steel wool in the bottom and one should be empty, as seen in Figure 1.20-1.

4. Examine the jars each day for one week and record your observations, noting such things as water level and appearance of the steel wool.

5. At the end of one week, study the recorded day-by-day observations and explain the noted changes.

Figure 1.20-1. Dishes with Inverted Jars, One with Steel Wool

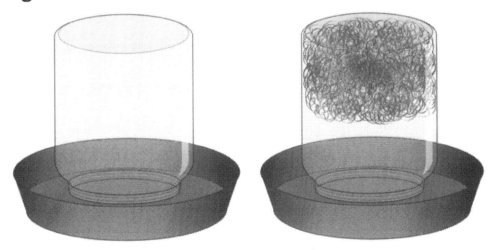

For Problem Solvers

Notice the color of the rust that forms on the steel wool. Try to find items that are made of steel or iron and that have been used for a long time and examine them for spots of rust. See whether you can find the same thing on items that are made of plastic. Find out what stainless steel is and then find some items that are made of stainless steel. Can you find rust on them? Look up stainless steel on the Internet or in your encyclopedia and try to find out how stainless steel is different and why it is used.

Teacher Information

As steel wool is exposed to moist air over a period of time, the moisture serves as a medium to bring oxygen molecules in the air in close contact with molecules of iron in the steel wool. Oxygen molecules and iron molecules combine to make iron oxide. This process uses up some of the oxygen in the air inside the jar, reducing the amount of gas (air) in the jar. This, in turn, reduces the air pressure inside the jar. Thus the atmospheric pressure outside the jar is greater than the pressure inside the jar, and water is forced into the jar. Student observations should include the rising water level inside the jar containing steel wool, as well as the rust color forming on the steel wool.

Integrating

Math, reading

Science Process Skills

Observing, inferring, comparing and contrasting, measuring

How Can Chemical Changes Help You Write a Secret Message?

(Take home and do with family and friends.)

Materials Needed
- Small jar
- Milk (only a few drops)
- Toothpick
- White paper
- Lamp with light bulb

Procedure
1. Dip the toothpick into the milk and use it as a pen to write a message on the paper. Let the milk dry.
2. What happens to your message as the milk dries?
3. Hold the paper close to a burning light bulb. What happens as the paper absorbs heat from the light bulb? What can you say about this?

For Problem Solvers
Try the same thing, using lemon juice instead of milk. What other things do you think might work? Try them.

Teacher Information
As the milk dries, the residue blends in with the white paper and becomes invisible. When heat is applied, a chemical reaction takes place in the milk residue, turning it dark and making it easily visible against the white paper.

Integrating
Language arts

Science Process Skills
Observing, inferring, predicting, comparing and contrasting

The Nature of Matter

How Does Temperature Affect the Speed of Molecules?

Materials Needed

- Two tumblers per small group
- Food coloring
- Paper and pencils
- Two eyedroppers per small group
- Hot and cold water
- "Science Investigation Journaling Notes" form for all students

Procedure

1. Do the following as a Science Investigation. Obtain a blank copy of the "Science Investigation Journaling Notes" from your teacher. Write your name, the date, and your question at the top. Plan your investigation through item 5 of the form (Procedure) and have it approved by your teacher. Complete the Journaling Notes as you perform your investigation. When you are finished, share your project with your group and submit your Journaling Notes to your teacher if requested.

2. Put very cold water in one tumbler and hot water in the other. Fill each about halfway full.

3. Draw four or five drops of food coloring into each of the two eyedroppers. Put as near the same amount in each dropper as possible.

4. Hold a dropper over each tumbler and squeeze to empty the contents of both at exactly the same time.

5. Compare the movement of the color in the two containers. In which tumbler did the color spread more rapidly?

6. If you have time, try different colors and different water temperatures. Record your observations.

For Problem Solvers

Obtain another copy of the "Science Investigation Journaling Notes" and try the same investigation, but using color as the variable this time. Put water of the same temperature in both containers. See whether one color of food coloring diffuses (mixes) through the water any faster than another color. Find a stopwatch and keep time to find out how long it takes to fully diffuse so that the water is equal in color throughout.

Next, test water at different temperatures, timing the diffusion at each temperature. Does temperature make a big difference, a small difference, or none at all?

Compare diffusion time of various liquids. Do you think the food coloring will spread through milk at the same rate as through water, if the two liquids are the same temperature? Try it. What other liquids could you compare?

Teacher Information

As temperatures increase, molecules move faster. The food coloring will diffuse noticeably more rapidly in hot water than in cold water. In this experiment, water temperature is the variable. Your "problem solvers" will try the same experiment with color as the variable. For instance, use two tumblers of cold water and put red in one and green (or blue) in the other. They are also encouraged to use a stopwatch and a thermometer and record the actual time required for maximum diffusion (equal color throughout, as judged by the students).

Integrating

Math

Science Process Skills

Observing, inferring, measuring, predicting, communicating, comparing and contrasting, using space-time relationships, formulating hypotheses, identifying and controlling variables, experimenting

How Does Temperature Affect Solubility?

(Teacher-supervised activity)

Materials Needed

- Two tumblers (equal size) per small group
- Cold water
- Hot water (or a heat source)
- Two spoons (or other stirring instruments) per small group
- Measuring spoons
- Sugar
- Markers
- "Science Investigation Journaling Notes" form for each student

Procedure

1. Do the following as a Science Investigation. Obtain a blank copy of the "Science Investigation Journaling Notes" from your teacher. Write your name, the date, and your question at the top. Plan your investigation through item 5 of the form (Procedure) and have it approved by your teacher. Complete the Journaling Notes as you perform your investigation. When you are finished, share your project with your group and submit your Journaling Notes to your teacher if requested.

2. Be sure the tumblers are equal size.

3. Make a mark on each tumbler about one-fourth of the way down from the top. The mark should be at exactly the same point on each tumbler.

4. Using cold water for one tumbler and hot water (not hot enough to burn you) for the other, fill each exactly to the mark.

5. Using a measuring spoon (a teaspoon is about right) put one level spoonful of sugar in each tumbler.

6. Stir the water in each tumbler until the sugar has completely dissolved in the water.

7. Which dissolved sugar faster, cold water or hot water?

8. Add another level spoonful of sugar and again stir until it is completely dissolved.

9. Continue doing this, counting the spoons full of sugar added to each tumbler. Stop adding sugar when you can no longer make it completely dissolve in the water.

10. Which dissolved more sugar, the cold water or the hot water? How much more? Why do you think this was so?

11. Complete your "Science Investigation Journaling Notes." Are you ready to explain how temperature affects solubility? Discuss it with your group.

For Problem Solvers

Do this project as a Science Investigation also. Obtain a blank copy of the "Science Investigation Journaling Notes." Write your name, the date, and your question at the top. Plan your investigation through item 5 (Procedure) and have it approved by your teacher. Complete the Journaling Notes as you perform your investigation. Share your project with your group, and submit your Journaling Notes to your teacher if requested.

Get some sugar cubes and see whether they dissolve at the same rate as loose sugar. Use a balance scale to see that you have the same amount of loose sugar as is in the sugar cube. Compare them in hot and cold water. Time the dissolving rate with a stopwatch. Measure the temperature of the hot water and the cold water. Considering the dissolving rate for both of these, predict the dissolving rate if the water is halfway between these two temperatures. Try it.

Do you think stirring has an effect on dissolving rate? See what you can do to find out.

Do you think salt dissolves at the same rate as sugar? Do you think water temperature has the same effect on the dissolving rate of salt as it has on sugar? Devise an experiment to find out, and carry out your experiment.

Teacher Information

Hot water molecules move more rapidly than cold water molecules do. The dissolving sugar molecules are therefore dispersed more completely throughout the liquid and more sugar can be dissolved in the hot water.

Your "problem solvers" will compare the dissolving rate of sugar cubes with the dissolving rate of loose sugar. They will also investigate the effect of stirring on the dissolving rate of sugar. If their interest holds, they will find out whether salt dissolves at the same rate as sugar. Your young scientists might think of other variables to test as well.

Caution: Care must be taken to be sure the hot water used is not hot enough to injure someone if spilled.

Integrating

Math

Science Process Skills

Observing, inferring, measuring, predicting, communicating, comparing and contrasting, using space-time relationships, formulating hypotheses, identifying and controlling variables, experimenting

How Can You Make Large Sugar Crystals from Tiny Ones?

(Teacher-supervised activity)

Materials Needed

- Drinking glasses or jars
- Water
- Cotton string or thread
- Pencils
- Sugar
- Pans
- Heat source
- Stirrers or wooden spoons

Procedure

1. Put a cup of water in the pan and heat it until it boils.
2. When the water begins to boil, turn off the heat and add about 1.5 cups of sugar and stir.
3. If all the sugar dissolves, add a bit more and stir. Keep doing this until no more sugar will dissolve in the water.
4. Let the water cool, and then pour it into a drinking glass.
5. Tie a piece of cotton thread or string to a pencil and lay the pencil across the glass, allowing the string to extend to the bottom of the glass.
6. Place the drinking glass on a shelf where it can remain undisturbed for several days.
7. Examine the contents of the glass, particularly the string, each day and write down your observations. DO NOT MOVE IT OR TOUCH ANY OF IT.
8. When you observe no more changes, try to explain what happened during the days the glass remained on the shelf.

The Nature of Matter

For Problem Solvers

Use the same procedure to experiment with other substances you can find. Include salt and powdered alum among the materials you try. Examine your crystals with a hand lens and with a microscope, if you have one. Observe and compare very carefully and share your observations with others.

Teacher Information

Caution: To avoid injury to students, boiled water should be handled only by the teacher.

As sugar dissolves in water, the crystals break down into molecules so small they cannot be seen, even with a powerful microscope. A molecule of sugar is the smallest particle of sugar that can exist. If it were any smaller, it would no longer be sugar.

In this activity, sugar crystals are dissolved into molecules, forming a supersaturated solution (containing more sugar in solution than could be dissolved at room temperature). Then, as the solution cools, crystals begin coming out of solution and collecting around the string. As this happens, large sugar crystals are formed. You could call it rock candy.

"Problem solvers" will also try forming crystals from other substances, such as salt and/or powdered alum, following the same procedure.

Integrating

Math, language arts

Science Process Skills

Observing, inferring, classifying, measuring, predicting, communicating, comparing and contrasting, using space-time relationships, formulating hypotheses, identifying and controlling variables, experimenting

What Does Litmus Paper Tell Us About Substances?

Materials Needed

- Red litmus paper
- Blue litmus paper
- Glass containing a small amount of vinegar water (about half vinegar and half water)
- Glass containing a small amount of baking soda mixed in water
- Glass containing a small amount of tap water
- Paper and pencils

Procedure

1. Write "Vinegar Water," "Baking Soda Water," and "Water" across the top of your paper.
2. Write "Blue Litmus Paper," "Red Litmus Paper," and "Acid, Base, or Neutral" down the left side of your paper, as shown in the sample on the next page.
3. Dip one end of a strip of blue litmus paper into the vinegar water.
4. Did it change color? If so, what color is it now?
5. Write the color in the space for vinegar water and blue litmus paper.
6. Dip one end of a strip of red litmus paper into the vinegar water.
7. What happened this time?
8. Write the color in the space for vinegar water and red litmus paper. If the litmus paper stayed the same color, write "no change."
9. Repeat steps 3 through 8 for the baking soda water.
10. Repeat steps 3 through 8 again for the water.
11. If a substance turns blue litmus paper red, we say the substance is an acid. If the substance turns red litmus paper blue, we say the substance is a base. If the substance does not change the color of either litmus paper, we say the substance is neutral.

12. Fill in the bottom line of your chart, identifying each of the three substances as either acid, base, or neutral.

13. Compare your information with others. Did they obtain the same results?

	Vinegar Water	Baking Soda Water	Water
Blue LP			
Red LP			
A, B, N			

For Problem Solvers

Stronger acids turn blue litmus paper darker red, and stronger bases turn red litmus paper darker blue. Find other substances to test with litmus paper. Some of the materials you could test are milk, tea, coffee, window cleaner, bathtub cleaner, and mouthwash. You will think of others as you go. Identify which of the materials you tested are strong acids, which are weak acids, which are strong bases, and which are weak bases.

Test a variety of brands of soft drinks. Before you test them, predict whether they will be acid, base, or neutral. If you think they will be acid or base, predict which drinks will be the strongest. After testing them, list them in the order of their strength, as shown by the litmus test.

Find as many sources of water as you can find in your area. These might include tap water, rain water, pond water, swamp water, river water, and others. Use the litmus paper test on each one, and list them in order according to your litmus test results.

Teacher Information

This activity will provide an introduction to the terms "acids" and "bases" and to the use of litmus paper as an indicator for determining which is which. It will also become a practical and useful experience for your problem solvers who decide to extend the activity into various drinks and into water from various sources. If acid rain is sometimes a problem in your area, you might also want to have your students collect samples of rain water during each storm and keep record of the acid levels from each. Do the same with snow; melt it down and test it.

As a long-term project, consider having your students determine whether the acidity of rain water changes throughout the year in your area.

In the absence of litmus paper, or in addition to it, try red cabbage juice. You can either boil the cabbage, then strain the juice to remove the solids, or put red cabbage and water in a blender, again straining the juice. To use it as an acid/base indicator, take a small amount of juice, such as a spoonful, and add a drop or two of the liquid being tested. Compare the color changes to the changes in litmus paper.

Integrating
Math, social studies

Science Process Skills
Observing, inferring, classifying, measuring, predicting, communicating, comparing and contrasting, experimenting

How Can Perfume Get into a Sealed Balloon?

Materials Needed

- Two balloons per group
- Two small bowls per group
- Perfume
- Water
- String

Procedure

1. Put one-half cup of water in each of the two bowls.
2. Mix several drops of perfume into the water of one bowl.
3. Blow up both balloons and tie them. Use a string with a bow knot so they can be untied later.
4. Place one balloon in each bowl. Press them into the bowl to create an air-tight seal. (See Figure 1.26–1.)
5. Leave the materials undisturbed for at least two hours.
6. After at least two hours, take the balloons to another room where the perfume from the bowl cannot be smelled.
7. Untie the balloon that was on the nonperfume bowl. Let the air out slowly and smell it.
8. Untie the balloon that was on the perfume bowl. Let the air out slowly and smell it.
9. What did you notice about the air in the balloons? What can you say about this?

For Problem Solvers

Do you think the perfume getting into the balloon would be affected by how tightly the balloon is blown up? Test this question by using three balloons. Be sure the balloons are identical, except that one will be blown up more and one less than before.

How about also testing the permeability of different plastic wraps? Put your perfume water in drinking cups or drinking glasses and seal plastic wrap over the tops of the containers; then see whether you can smell the

Figure 1.26-1. Balloons in Bowls

perfume through the plastic. Compare different brands of plastic wrap, and perhaps even different brands of perfume. Before you begin, make your predictions. Will they be the same? Do you think permeability of plastic wraps matters with foods that are stored in the refrigerator? Why?

Teacher Information

Molecules in the perfume are small enough to permeate the wall of the balloon. When the air is let out of the balloon after a two-hour period, the perfume in the air of the balloon should be evident from the smell.

Foods sometimes take on odors from each other while wrapped and in the refrigerator. Your "problem solvers" will find out why.

Integrating

Math

Science Process Skills

Observing, inferring, classifying, measuring, predicting, communicating, comparing and contrasting, using space-time relationships, formulating hypotheses, identifying and controlling variables, experimenting

How Can You Cause Molecules to Move Through Solids?

Materials Needed
- Balloons
- String
- Markers
- Paper and pencils

Procedure
1. Blow up a balloon and tie it.
2. Measure the size of the balloon by wrapping the string around it at the largest point and marking the string. Record the length of string required to go around the balloon.
3. Place the balloon where it will not be disturbed and where the temperature will remain quite constant.
4. For three days, measure the balloon twice a day with the same string and mark the string to indicate the length required to go around the balloon. Each time you measure, record the length of string required.
5. At the end of three days, describe your observations. Try to explain any changes you noted.

For Problem Solvers

Do this project as a Science Investigation. Obtain a blank copy of the "Science Investigation Journaling Notes." Write your name, the date, and your question at the top. Plan your investigation through item 5 (Procedure) and have it approved by your teacher. Complete the Journaling Notes as you perform your investigation. Share your project with your group, and submit your Journaling Notes to your teacher if requested.

Find some balloons of different brands and different quality and repeat this activity. Set up an experiment to compare the different types of balloons. Be sure to use balloons of the same size and shape, to blow them up to the same size, and to tie them in the same way.

Teacher Information

You might check to see that balloons are tied tightly so air cannot leak through the opening. You can do this by submerging them in water to check for air bubbles. As the balloons sit, air molecules actually permeate the balloon walls and they will lose air slowly, even though air is not escaping by any observable means. For the duration of this activity, the air temperature should remain as constant as possible. If air temperature changes, balloons will expand or contract (in warmer and cooler air, respectively), which will reduce the reliability of the results.

Integrating

Math

Science Process Skills

Observing, inferring, classifying, measuring, predicting, communicating, comparing and contrasting, using space-time relationships, formulating hypotheses, identifying and controlling variables, experimenting

What Is Viscosity?

Materials Needed

- Four tall olive jars with lids (or other tall, skinny jars) per small group
- Four marbles (different colors) per small group
- Corn syrup
- Mineral oil
- Vegetable oil
- Water
- Paper and pencils

Procedure

1. Be sure all four jars are the same size.
2. Place a marble in each jar.
3. Fill each jar with one of the liquids and put the lid on it. There should be no air under the lid.
4. When all lids are tightly in place, get someone to help you turn all four jars upside down at once. Observe the marbles.
5. Record which marble sank to the bottom first, second, third, and fourth. Repeat and compare the results with your first trial.
6. Test other liquids and compare with these.
7. Discuss your findings with your friends or your teacher.

For Problem Solvers

Read about viscosity in a dictionary. Find an encyclopedia article that tells about the viscosity of oil and read the article. Look up viscosity on the Internet. Ask a mechanic why oil is made at different viscosities for automobile engines. Why does it matter, and what are the advantages of light oil (low viscosity) and of heavy oil (high viscosity)? Some engine oils are even multiple viscosity. What does that mean, and why do they make them that way?

Teacher Information

Other liquids can be substituted for those listed above, but they should vary in viscosity (thickness). The marbles will sink more slowly in liquids with greater viscosity. Viscosity is resistance to flow.

If olive jars or other tall, thin jars are not available, baby food jars can be used. Try to get the larger size, for height. Test tubes work very well, if they are available. They must have stoppers, of course. Sturdy test tubes with threaded caps are sold as mini-bottles, baby soda bottles, and by other names. These are ideal for this activity and are available from science supply companies. They are actually the blanks for two-liter bottles—before they are transformed to their grown-up size.

The activity can even be done using open bowls. Put the liquids in separate bowls and a spoon in each bowl. Students should take a spoonful of the liquid and pour it back into the same bowl, observing how fast it pours out of the spoon. This doesn't have quite the interest or accuracy of the marble activity, but it will work.

As a practical application of this concept, your "problem solvers" will find out why automobile engines use oils of various viscosities. One factor is temperature. As is true of honey, oils become thinner (less viscous) as they become warmer. Heavier oil is generally preferred for hot weather and thinner oil for cold weather. Modern oils are also made in multiple viscosities; they have properties that cause them to behave as a heavier oil in hotter temperatures and as a lighter oil in colder temperatures.

Integrating

Reading, math

Science Process Skills

Observing, inferring, classifying, measuring, predicting, communicating, comparing and contrasting, using space-time relationships, formulating hypotheses, identifying and controlling variables, experimenting, researching

🪑 Can You Solve This Nature of Matter Word Search?

Try to find the following Nature of Matter terms in the grid below. They could appear in horizontal (left to right), vertical (up or down), or diagonal (upward or downward) position.

dissolve	solution	mixture
gas	condensation	molecules
viscosity	chemical	reaction
states	liquid	solid

```
R  E  A  C  T  I  O  N  Q  W  E  R
U  Y  V  C  H  E  M  I  C  A  L  T
I  N  M  A  L  K  J  H  S  G  F  D
O  B  D  N  P  G  F  E  D  S  A  S
P  D  I  S  S  O  L  V  E  T  W  A
L  V  U  J  H  U  R  U  Y  R  E  Q
K  C  Q  K  C  U  G  A  S  T  D  W
J  X  I  E  L  P  O  I  T  U  I  E
S  O  L  U  T  I  O  N  A  I  L  R
C  O  N  D  E  N  S  A  T  I  O  N
M  Z  M  I  X  T  U  R  E  B  S  N
H  G  F  V  I  S  C  O  S  I  T  Y
```

Can You Create a New Nature of Matter Word Search of Your Own?

Write your Nature of Matter words in the grid below. Arrange them in the grid so they appear in horizontal (left to right), vertical (up or down), or diagonal (upward or downward) position. Fill in the blank boxes with other letters. Trade your Word Search with someone else who has created one of his or her own, and see whether you can solve the new puzzle.

_____   _____   _____

_____   _____   _____

_____   _____   _____

_____   _____   _____

_____   _____   _____

The Nature of Matter   **69**

Answer Key for Nature of Matter Word Search

```
R E A C T I O N Q W E R
U Y V C H E M I C A L T
I N M A L K J H S G F D
O B D N P G F E D S A S
P D I S S O L V E T W A
L V U J H U R U Y R E Q
K C Q K C U G A S T D W
J X I E L P O I T U I E
S O L U T I O N A I L R
C O N D E N S A T I O N
M Z M I X T U R E B S N
H G F V I S C O S I T Y
```

Hands-On Physical Science Activities

Do You Recall?

Section One: The Nature of Matter

1. Explain why a paper clip can float on water.

2. Why does a dry sponge weigh less than when it was wet?

3. When the grass is dry at night and wet in the morning, but there was no rain and no sprinklers, where did the water that is on the grass come from?

4. What is a mixture?

5. What is a solution?

Do You Recall? *(Cont'd.)*

6. Suppose you were to swim in your local swimming pool, then in a pool of salt water. In which of these would it be easier for you to float on the surface? Why is that?

7. Consider a submarine that contains a large tank. The tank can be filled with water from the sea or the water can be pumped out, leaving the tank empty or partly full. Explain how this can be used to change the depth of the submarine at any given time.

8. What common substance exists in nature in all three states?

9. If a burning candle is lowered into a jar of carbon dioxide, will the flame go out, burn more brightly, or continue burning the same as before?

10. How does a raw egg respond different from a boiled egg when both are placed in a large cup of water and salt is added slowly? Why is that?

Do You Recall? *(Cont'd.)*

11. When wood burns, is it a physical change or a chemical change?

12. When an iron nail rusts, is it a physical change or chemical change?

13. If food coloring is placed in a cup of hot water and in a cup of cold water, will the food coloring disperse through the water:
    a. More rapidly in the hot water?
    b. More rapidly in the cold water?
    c. Equally in both?

14. If salt is placed in a cup of hot water and in a cup of cold water, will the salt dissolve in the water:
    a. More rapidly in the hot water?
    b. More rapidly in the cold water?
    c. Equally in both?

15. When litmus paper is dipped into a liquid, what kind of information does it provide?

Answer Key for Do You Recall?

Section One: The Nature of Matter

Answer	Related Activities
1. Water molecules hold to each other (cohesion).	1.3
2. Water has weight; after the water evaporates, there is less weight.	1.4
3. From the air	1.5
4. Two or more substances mixed together but both retain their separate identities	1.6
5. A substance placed in a liquid seems to become a part of the liquid.	1.6
6. In the pool of salt water	1.8, 1.9
7. More water in the tank increases the density of the submarine, and the submarine moves to a deeper level.	1.10
8. Solid, liquid, and gas; water	1.11
9. The candle will go out. It must have oxygen to burn.	1.13
10. The raw egg will rise first. It is less dense than the boiled egg.	1.8
11. Chemical change; a new substance is formed and it is not reversible	1.19
12. Chemical change	1.20
13. a. More rapidly in the hot water	1.22
14. a. More rapidly in the hot water	1.23
15. Whether the liquid is an acid, a base, or neutral	1.25

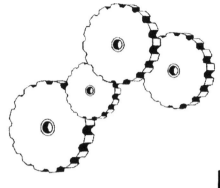

Energy

To the Teacher

We use many forms of energy every day, yet we never see it. The sun's energy literally powers the earth, but it is so common we take it for granted. It is just there. Energy comes in many forms, none of which looks like "energy." It is disguised as a matchstick, a lump of coal, a gallon of gasoline, or a glass of orange juice; it is never just energy. In a broad sense, energy is so much a part of us and our surroundings that it would be impossible to deal with it as a topic separate from other topics treated in this book. The sun's energy keeps us warm and gives us light. Part of the sun's energy is converted by plants into food for people and other animals. Cattle convert some of that energy from plant form to muscle, which we eat (beefsteak, hamburger, and so forth). Energy is consumed by people in both plant and animal form. We, in turn, convert it into human flesh and bones. The sun's energy is a vital ingredient of our own bodies and of a great deal of our surroundings. As energy is transformed from one form to another, it is never lost or destroyed. One of the basic principles of science is "conservation of energy." There is always the same total amount. Electrical energy can be transformed to mechanical energy, and various forms of energy can be transformed to various other forms of energy, but the total amount of energy in the universe does not change.

In a narrower sense, energy is sometimes defined as the capacity to perform work (*Webster's New Collegiate Dictionary*). It exists in two forms, *potential* energy and *kinetic* energy. Potential energy is the ability to do work. Work is defined as force acting through a distance. Specifically, Work = Force times Distance. Kinetic energy is the energy of motion. A rubber band stretched out has potential energy. When it is released, its potential energy is converted to kinetic energy. A stick of dynamite has potential energy. When a small electrical charge or the right amount of heat is applied, the potential energy is converted to kinetic energy with great force.

The potential energy in dynamite is chemical energy. The potential energy in the rubber band, in a set mousetrap, or in a raised hammer is mechanical energy.

The energy we consume in the form of meat, fruit, and vegetables isn't all used to build body cells. We use some of it to walk and talk. We even use some of this energy as we think.

Most sections of this book deal directly or indirectly with forms of energy. This section recognizes these but emphasizes additional topics, such as heat, gravity, and the relationship between energy and work.

Activities that appear complex can usually be used in the lower grades by de-emphasizing terminology and mathematical applications. With simplified explanations of the concepts, children can participate in the activities and benefit from exposure to these principles by increasing their readiness for future learning in greater depth.

The following activities are designed as discovery activities that students can usually perform quite independently. You are encouraged to provide students (usually in small groups) with the materials listed and a copy of the activity from the beginning through the "Procedure." The section titled "Teacher Information" is not intended for student use, but rather to assist you with discussion following the hands-on activity, as students share their observations. Discussion of conceptual information prior to completing the hands-on activity can interfere with the discovery process.

Regarding the Early Grades

With verbal instructions and slight modifications, many of these activities can be used with kindergarten, first-grade, and second-grade students. In some activities, steps that involve procedures that go beyond the level of the child can simply be omitted and yet offer the child an experience that plants the seed for a concept that will germinate and grow later on.

Teachers of the early grades will probably choose to bypass many of the "For Problem Solvers" sections. That's okay. These sections are provided for those who are especially motivated and want to go beyond the investigation provided by the activity outlined. Use the outlined activities and enjoy worthwhile learning experiences together with your young students. Also consider, however, that many of the "For Problem Solvers" sections can be used appropriately with young children as group activities or as demonstrations, still giving students the advantage of an exposure to the experience and laying groundwork for connections that will be made later on.

Correlation with National Standards

The following elements of the National Standards are reflected in the activities of this section.

K-4   Content Standard A: Science as Inquiry

As a result of activities in grades K-4, all students should develop:

1. Abilities necessary to do scientific inquiry
2. Understanding about scientific inquiry

K-4   Content Standard B: Physical Science

As a result of activities in grades K-4, all students should develop understanding of

1. Properties of objects and materials
2. Position and motion of objects
3. Light, heat, electricity, and magnetism

5-8   Content Standard A: Science as Inquiry

As a result of activities in grades 5-8, all students should develop:

1. Abilities necessary to do scientific inquiry
2. Understanding about scientific inquiry

5-8   Content Standard B: Physical Science

As a result of activities in grades 5-8, all students should develop understanding of

1. Properties and changes of properties in matter
2. Motions and forces
3. Transfer of energy

In What Ways Is Energy Used in Our Neighborhood?

Materials Needed

- Paper (divided into two columns)
- Pencils

Procedure

1. Divide your paper in half with a vertical line about 21 cm (4¼ inches) from each side.
2. Go for a walk around your neighborhood, looking for indications that energy is being used.
3. In the left column of your paper, make a note of anything you see happening that uses some form of energy.
4. Decide what you think is the form of energy being used, and write that in the right-hand column.
5. Discuss your observations with the class.

Teacher Information

Student observations should include lights burning, fans running, cars moving on the road, or any other event that uses some form of energy. For each of these, students should write in the right-hand column the form of energy they think is being used (electricity, gasoline, food, and so forth). Emphasize that if something moves, energy is being used, even if it is only a paper or a leaf being blown by the wind. What about the chemical energy used by students as they write, talk, and walk?

Integrating

Language arts, social studies

Science Process Skills

Observing, inferring, math, communicating

How Well Do You Conserve Energy at Home?

(Take home and do with family.)

Materials Needed

- Pencils and paper

Procedure

1. Are you careful about how much energy you use? Use the following energy checklist at home to see how well you and others who live there are conserving energy. Perhaps you would like to add other items to the list.

   ☐ Is cold water or warm water used when washing clothes, instead of hot water?

   ☐ Are draperies used to increase efficiency of heating and air conditioning by opening them when sunlight needs to be let in and closing them at times when they are needed as insulation?

   ☐ Are air conditioners and furnaces used sparingly?

   ☐ Are heating and cooling systems serviced regularly to assure safe and efficient operation?

   ☐ Are filters in the heating and cooling systems regularly serviced (cleaned or changed)?

   ☐ Are heating and cooling vents free from obstruction?

   ☐ If there is a fireplace, is the damper on the chimney closed when the fireplace is not in use?

   ☐ Are lights used only when needed?

   ☐ Are electrical appliances turned off when they are not in use?

   ☐ Are refrigerator doors closed promptly after use?

   ☐ Are refrigerator door gaskets tight, so they seal properly?

   ☐ Are windows and doors tight, and do they prevent drafts? Is caulking and weather stripping around doors and windows in good condition?

2. After completing the home energy checklist, compare notes with others and discuss things that need to be done in your homes to help conserve energy.

For Problem Solvers

Decide which parts of the energy checklist also apply at school. Using this list of things to check, or another list that you write, find out how well energy is being conserved at school. You might have some suggestions for improvement. You could do this activity as a group if you prefer. Consider writing a letter to the school principal, custodian, or to the school district office, outlining your findings and suggestions. In most cases, the principal and custodian would probably be the best people to notify of your concerns.

Teacher Information

This activity should help your students to be more energy conscious. If it seems appropriate, suggestions from the class could be sent out to the homes, listing only general ideas, and not pointing out individual problems which might be offensive.

If the energy survey at school produces fruitful recommendations, invite the media to feature the class with a short story spotlighting the efforts of the students. Do this only if it does not create embarrassment for others involved.

Integrating

Language arts, math, social studies

Science Process Skills

Observing, inferring, classifying, measuring, predicting, communicating, using space-time relationships, formulating hypotheses, identifying and controlling variables, experimenting, researching

How Do Potential Energy and Kinetic Energy Compare?

(For older children or teacher demonstration)

Materials Needed

- Mousetrap
- Ball
- String
- Pencil eraser

Procedure

1. Set the mousetrap. (See Figure 2.3-1.) What kind of energy does it have? (Notice that the trap is loaded and ready to snap closed as soon as the release lever is triggered.)

Figure 2.3-1. Mousetrap

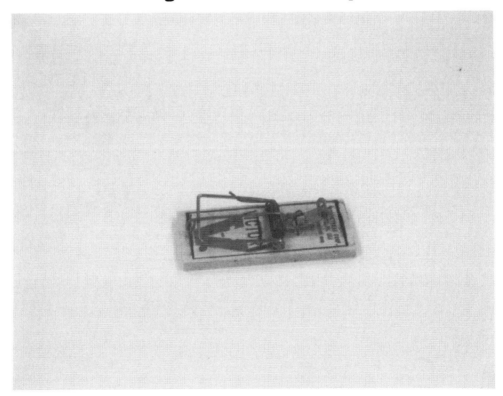

2. Drop the pencil eraser on the release lever of the mousetrap. (See Figure 2.3-2.) What kind of energy is present at the moment the trap springs shut?

3. Drop the ball onto a table and let it bounce a few times. (See Figure 2.3-3.) Describe one full bounce, from the top of the bounce to the bottom and back to the top, in terms of the presence of kinetic and potential energy.

4. Tie the eraser to the string. (See Figure 2.3-4.) Hold the end of the string and swing the eraser like a pendulum. Describe one full swing, back and forth, in terms of the presence of kinetic and potential energy.

For Problem Solvers

Think through the actions involved in a baseball game. List as many examples of potential energy as you can, and list as many examples of kinetic energy as you can. What about the pitcher standing in a relaxed position with ball in hand? The moment when the pitcher is wound up and ready to release the ball? The moment the ball connects with the bat? When the ball is zooming toward left field and the left-fielder is jumping to catch it? Add all of the actions you can think of, and list them as examples of either potential energy or kinetic energy.

Figure 2.3-2. Eraser Ready to Drop on Mousetrap

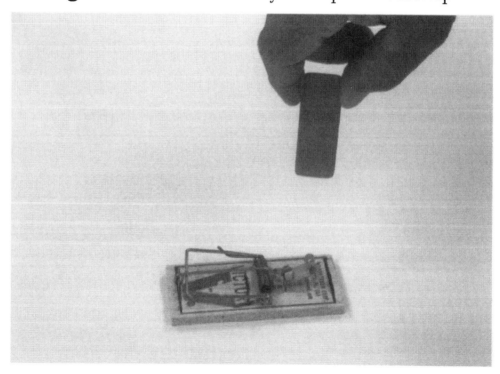

Figure 2.3-3. Ball Ready to Drop on Table

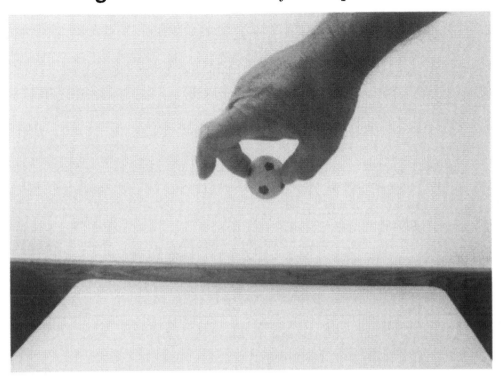

Figure 2.3-4. Eraser Pendulum

Teacher Information

This activity requires the student to distinguish between kinetic and potential energy, with the assumption that students have already had an introduction to those types of energy. When the mousetrap is set, the presence of the loaded spring gives it potential energy. At the moment the spring is released, the potential energy is converted to kinetic energy, or moving energy. Great care should be taken to assure that fingers are not injured by the mousetrap.

As the ball is held above the table, it has potential energy. The ball is released and the potential energy is converted to kinetic energy as the ball falls toward the table. When the ball strikes the table, it is compressed and the kinetic energy is converted to potential energy. The potential energy in the compressed rubber propels the ball into the air, again converting the potential energy to kinetic energy. Some of the energy escapes in the form of heat as the ball meets resistance with the air and the table. Thus, the height of the cycle decreases as the ball bounces.

The swinging pendulum passes through a cycle similar to that of the bouncing ball. All energy contained in the system is potential energy at the instant the pendulum is all the way to the top on either side of the cycle. At the moment the pendulum is at the bottom, moving neither upward nor downward, all of the energy is kinetic.

If the pendulum is at rest, it has neither potential nor kinetic energy. When in a position that gravity can cause movement when released, it has potential energy.

Your baseball stars and fans will enjoy dissecting a baseball game in terms of what they have learned about potential energy and kinetic energy and identifying the changes from one form of energy to the other.

Integrating

Math

Science Process Skills

Observing, inferring, classifying, measuring, predicting, communicating, comparing and contrasting, using space-time relationships, formulating hypotheses, identifying and controlling variables, experimenting

Hands-On Physical Science Activities

How Is Work Measured?

Materials Needed

- One-pound weights
- Foot rulers

Procedure

1. Stand the ruler on the table or the floor.

2. Raise the one-pound weight to the top of the ruler. The amount of work you did to raise one pound a distance of one foot is called one foot-pound.

3. Raise the one-pound weight six inches. How much work did you do? How much potential energy does the weight have at that point?

4. Raise the weight two feet. How much work did you do this time? How much potential energy does the weight have?

5. Set the weight on the table. How much potential energy does it have now?

6. Slide the weight to the edge of the table. How much potential energy does it have at that position?

7. Try to determine the amount of potential energy of various objects from different positions and the amount of work required to move those objects certain distances.

8. Climb a flight of stairs. How much work did you do to get to the top? What is your potential energy, assuming the possibility of falling or jumping to the bottom?

For Problem Solvers

Find a ten-pound weight. Lift the weight over your head. Measure the distance from the floor to the highest point you lifted the weight. Measure the height, in feet, and multiply that number by ten. That's the number of foot-pounds of work you did in lifting the weight one time. How much work can you do with the same weight in one minute? Do the same exercise every other day for two weeks. Make a graph showing the amount of work you do with the weight each time. It's okay to practice between your measured sessions if you want to. See how much you can improve in two weeks. Maybe your friends would like to join you and see how much they can improve, too.

Teacher Information

A durable plastic or cloth bag filled with one pound of sand, beans, or other material would be an excellent weight for this activity. If a hard object is used, several thicknesses of newspaper or other padding could be placed on the table or floor to muffle the sound when the weight is dropped and to protect the surface from possible damage.

The foot-pound is a standard unit for measuring work or potential energy. (The metric system equivalent to foot-pounds is Newton-meters, or joules, not at all in common usage in the United States.)

Integrating

Math

Science Process Skills

Observing, inferring, measuring, predicting, communicating, using space-time relationships, identifying and controlling variables, experimenting

How Can Wind Energy Be Used to Turn Something?

Materials Needed

- Square poster paper or card stock (about 15 cm [6 in.] square)
- Round pencils with erasers
- Straight pins
- Scissors
- Rulers
- Staplers

Procedure

1. Draw two lines on a card from corner to corner. The lines should cross at the center.
2. Make a pencil mark at the center of the card, where the lines cross. (See Figure 2.5–1.)

Figure 2.5–1. Card Prepared for Making a Pinwheel

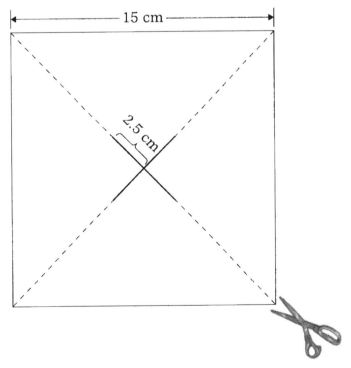

3. You now have a center point and four lines that connect the center point to the corners of the card.

4. Make a pencil mark on each of the four lines, 2.5 cm (1 in.) from the center point, as shown in Figure 2.5-1.

5. Make four cuts in the card with the scissors. Cut along each line, from the corner toward the center point, to the mark you made on the line.

6. You now have eight points. Fold every other point into the center of the card.

7. Staple the folded points at the center. This is your pinwheel.

8. Push the straight pin through the center of the pinwheel and into the eraser of the pencil. (See Figure 2.5-2.) Blow on the pinwheel to see that it spins freely.

9. When the wind blows, take your pinwheel outdoors. Share it with others.

Figure 2.5-2. Pinwheel

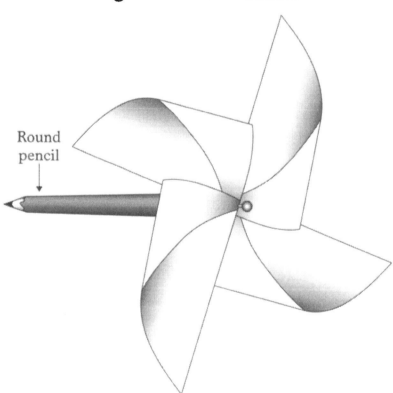

Round pencil

Hands-On Physical Science Activities

For Problem Solvers

Find two plastic or paper drinking cups, tape, and some string. One of the cups can be very small. Push the pencil through the larger cup, so the cup becomes a stand for the pinwheel. Tape one end of the string to the pencil, near the end opposite the pinwheel, and fasten the other end of the string to the small cup. (See Figure 2.5–3.) Blow on the pinwheel and the string should wind around the pencil, lifting the cup. Now let's call it a windmill.

Next, your challenge is to think of other ways to move things with wind power. You might make a different design of windmill or a completely different device. Discuss your ideas with others and learn together.

By the way, where does wind energy come from? Trace its source.

Figure 2.5-3. Windmill Ready to Raise a Load

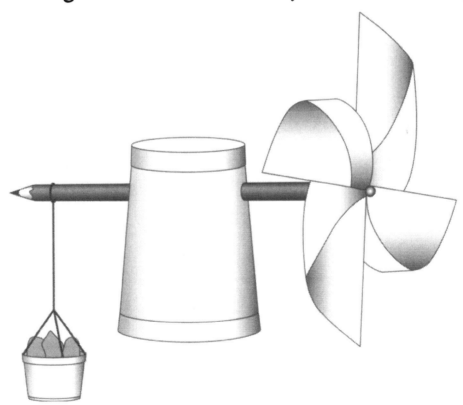

Teacher Information

In this activity, students use wind as a source of energy. Discuss the notion that even wind energy comes from the sun. Challenge students to explain that.

As students create objects that use wind as a source of energy to move something, they develop their creativity and problem-solving skills. If your problem solvers get bogged down in trying to think of ideas, you might want to encourage them to first research ways that wind power has been used and is now used. As they learn about ways wind power has been used historically, perhaps new and creative ideas will come to them. Encourage students also to discuss their ideas with one another. Talking about an idea often stimulates new thoughts for the person who is doing the explaining, as well as for the listener.

Integrating

Reading, language arts, math, social studies

Science Process Skills

Observing, inferring, measuring, predicting, communicating, identifying and controlling variables, experimenting, researching

Hands-On Physical Science Activities

How Can the Energy of Sound Cause Something to Move?

Materials Needed

- Two guitars
- Copies of "Science Investigation Journaling Notes" for all students

Procedure

1. Do the following as a Science Investigation. Obtain a blank copy of the "Science Investigation Journaling Notes." Write your name, the date, and your question at the top. Plan your investigation through item 5 of the form (Procedure) and have it approved by your teacher. Complete the Journaling Notes as you perform your investigation. When you are finished, share your project with your group, and submit your Journaling Notes to your teacher if requested.

2. Be sure the two guitars are tuned alike.

3. Stand the two guitars face-to-face, about 5 to 10 cm (2 to 4 in.) apart.

4. Strum the strings of one guitar. After two or three seconds, silence the strings of the guitar you strummed by putting your hand on them.

5. Listen carefully to the other guitar.

6. What do you hear? How did it happen?

7. Complete your "Science Investigation Journaling Notes." Are you ready to explain how the energy of sound can cause something to move? Discuss it with your group.

Teacher Information

This activity shows that sound can actually do work. It can make something move. Energy is transferred from one guitar to the other by sound waves, and the strings of the second guitar vibrate. The two guitars should be tuned alike so the vibrating frequency is the same for the two sets of strings.

Integrating

Music

Science Process Skills

Observing, inferring

How Can Magnetism Do Work?

Materials Needed

- Magnets
- Steel balls

Procedure

1. Place a magnet on the table.
2. Place a steel ball on the table about 2 to 3 cm (1 in.) from the end of the magnet.
3. Let go of the steel ball.
4. What happened?
5. What is work, and how was work done in step 3?

For Problem Solvers

Do this project as a Science Investigation. Obtain a blank copy of the "Science Investigation Journaling Notes." Write your name, the date, and your question at the top. Plan your investigation through item 5 (Procedure) and have it approved by your teacher. Complete the Journaling Notes as you perform your investigation. Share your project with your group, and submit your Journaling Notes to your teacher if requested.

Find a variety of magnets. Predict which ones are strongest and weakest, and lay them out in order from strongest to weakest, according to your predictions. Then continue with the above activity, comparing the strength of these magnets. Which one seems to attract the steel ball from the farthest distance?

Were your predictions accurate? Compare size with strength. Are larger magnets always stronger than smaller magnets?

Teacher Information

Work is defined in "To the Teacher," at the beginning of this section, as moving something (a force acting through a distance). The magnet should cause the steel ball to roll toward it. If this did not happen, try putting the steel ball a bit closer to the magnet or find a stronger magnet.

A paper clip can be used in place of the steel ball if necessary.

Integrating

Math

Science Process Skills

Observing, inferring, classifying, measuring, predicting, communicating, formulating hypotheses, experimenting

How Much Energy Is Stored in a Bow?

(Teacher-supervised activity to be done outdoors)

Materials Needed

- Toy bows
- Toy arrows tipped with suction cups
- Spring scales
- Foot rulers
- Measuring tape

Procedure

1. Do this activity outdoors. Find an isolated area.
2. Put the arrow on the bow. Hold the bow and arrow at a comfortable height and point the arrow straight ahead in a direction away from people.
3. Draw the bowstring back six inches and let it go.
4. Measure the distance the arrow traveled.
5. Attach the spring scale to the bowstring and pull the string back six inches. How many ounces or pounds of force were required to pull the string back six inches?
6. Predict the amount of force required to pull the string back one foot. Measure it with the spring scale.
7. Predict the distance the arrow will travel with the string pulled back one foot.
8. Shoot the arrow with the string pulled back one foot. Be sure the bow is held at the same height as before and still aimed straight ahead.
9. Measure the distance and compare with your predictions.
10. Predict the force required to pull the string back 1.5 feet and the distance the arrow will travel. Try it and test your predictions.

For Problem Solvers

Try a similar test with your own throwing ability. Find (or prepare) a one-kilogram (2.5-pound) weight. With your arm fully outstretched, pull your arm back 10 cm (4 in.), and from that point see how far you can throw the weight. For this exercise, hold your body still and make your arm do the work. Have a friend measure the distance your arm moves. Try it again with a 20-cm (8-in.) throwing range, then 30-cm (12-in.), and so on. Find your optimum throwing position (the position at which you can throw the farthest).

Where did the energy come from that shot the arrow? What about the energy that threw the weight? Trace the energy to its original source.

Teacher Information

Even though the arrow used in this activity is tipped with a suction cup for safety, close supervision is very important. Injury can still result if a child is hit in the face with the arrow. Young children can do the activity with less measuring and still predict the distances the arrow will travel. A ruler (or stick) could be marked at appropriate points to indicate the distance from the string to the bow. Children can indicate their predictions for distance the arrow will travel by placing a marker on the ground.

Integrating

Math, physical education

Science Process Skills

Observing, inferring, classifying, measuring, predicting, communicating, comparing and contrasting, using space-time relationships, formulating hypotheses, identifying and controlling variables, experimenting

How Can You Power a Racer with a Rubber Band?

(Take home and do with family and friends.)

Materials Needed

- Rubber band
- Thread spool
- Cotton swab (or sandwich skewer)
- Washer
- Paper clip

Procedure

1. Thread the rubber band through the spool.
2. Put the paper clip through one end of the rubber band to prevent the rubber band from slipping back through the hole in the spool.
3. Thread the other end of the rubber band through the washer.

Spool Racer

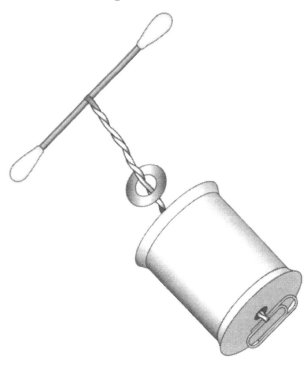

Hands-On Physical Science Activities

4. Insert the cotton swab through the rubber band, next to the washer.

5. Position the cotton swab so that one end is farther than the other from the rubber band.

6. Turn the cotton swab around and around to wind up the rubber band.

7. Set your racer on the floor and let it go!

8. Have races with others who made a similar racer.

9. What provides the energy for your racer? Is it really the rubber band? Think about it and discuss your ideas with others.

For Problem Solvers

Find a shorter rubber band. Predict whether your racer will do better with the shorter rubber band, or not as well. Try a longer rubber band, but be sure to predict again before you test it. Find the best rubber band for the racer.

Could you make a similar racer with other cylinders? How about a small soup can? A larger can? A two-liter plastic bottle? Other things? Be creative. What else can you design that would be powered by a rubber band? For larger cylinders, what can you use in place of the cotton swab?

Teacher Information

This is a good opportunity for students to work with a familiar object as a source of energy and to trace energy sources. It's also a good opportunity for children to explore creatively and to share their ideas. As they work together, they will learn about the benefits of cooperative efforts.

Students should recognize that the rubber band has no energy until energy is put into it by the winding action. The winding energy is fed into the rubber band by muscles of the human body. The muscles acquire their energy, directly or indirectly, from food made by plants—out of nutrients from the ground and the energy of the sun.

Integrating

Math

Science Process Skills

Observing, inferring, measuring, predicting, communicating, comparing and contrasting, identifying and controlling variables, experimenting

Energy

How Does a Nail Change as It Is Driven into a Board?

(Teacher-supervised activity)

Materials Needed

- Hammers
- Nails, at least 5 cm (2 in.) long
- Boards, at least 4 cm (1½ in.) thick
- Pounding surface
- A copy of the "Science Investigation Journaling Notes" for this activity for each student

Procedure

1. As you complete this activity, you will keep record of what you do, just as scientists do. Use a copy of the form "Science Investigation Journaling Notes" and write the information that is called for, including your name and the date.

2. For this activity you will learn about changes that occur in a nail as it is driven into a board. For item 1, the question is provided for you on the form.

3. Item 2 asks for what you already know about the question. If you have some ideas about what change might occur to the nail, write your ideas.

4. For item 3, write a statement of how a nail changes when it is driven into a board, according to what you already know, and that will be your hypothesis.

5. Now continue with the following instructions. Complete your Journaling Notes as you go. Steps 6 through 11 below will help you with the information you need to write on the form for items 4 and 5.

6. Place the board on a good pounding surface such as another board, a stack of newspapers, or concrete.

7. Pound the nail at least 2.5 cm (1 in.) into the board. Do not pound it all the way in.

8. As soon as you stop pounding, feel the nail. What difference do you notice in the nail?

9. Pull the nail out of the board with the hammer.

10. As soon as you get the nail out of the board, feel it again.

11. What difference do you notice in the nail by feeling it? What can you say about this?

12. Complete your "Science Investigation Journaling Notes." Are you ready to explain how a nail changes as it is driven into a board? Discuss it with your group.

For Problem Solvers

Rub your hands together, hard and fast. Do you feel a temperature change? How is this similar to what you experienced with the nail in the above activity?

Teacher Information

As the nail is pounded into the board, some of the energy from the moving hammer is changed to heat energy due to friction between the nail and the board. As the nail is removed from the board, friction again changes some of the energy to heat. If the nail is pulled out quickly, the heat might be even more noticeable than when it was pounded in.

Integrating

Math, language arts

Science Process Skills

Observing, inferring, comparing and contrasting

Science Investigation

Journaling Notes for Activity 2.10

1. Question: *How does a nail change as it is driven into a board?*

2. What we already know:

_____

_____

3. Hypothesis:

_____

_____

4. Materials needed:

_____

_____

5. Procedure:

_____

_____

6. Observations/New information:

_____

_____

7. Conclusion:

_____

_____

How Do Molecules Behave When Heated?

(Teacher demonstration)

Materials Needed

- Chalk or masking tape

Procedure

1. Have several students stand in a group.
2. Mark a border on the floor around the group with chalk or tape. Leave a few inches between the group and the border all the way around.
3. Ask students to move around slowly. Everyone should move constantly, but no one should move fast and there should be no pushing and shoving. They are to try to stay within the border marked on the floor.
4. Now instruct those in the group to move a bit faster. They are still to try to stay within the border.
5. Continue speeding up the movement of the group until they can no longer remain within the line marked on the floor.
6. Discuss what happened as those in the group increased their speed. Discuss how this relates to the movement of molecules as temperature is increased.

Teacher Information

As a substance increases in temperature, molecules do not increase in size, but they move faster. In their rapid movement, they bump into each other more frequently and with greater energy, and they require more space, just as did the group of students. If the members of the group were all running at top speed, they would have required much more space.

Integrating

Physical education

Science Process Skills

Observing, inferring, comparing and contrasting

What Happens to Solids as They Are Heated and Cooled?

(Teacher-supervised activity)

Materials Needed

- Wire, about 1 m (1 yd.) long
- Large nails or small bolts
- Candles
- Matches

Procedure

1. Wrap one end of the wire around the nail and anchor the other end to a support. Adjust the wire so the nail swings freely and barely misses the table or floor.
2. Light the candle and heat the wire.
3. Observe the nail. What happened?
4. Remove the candle and allow the wire to cool.
5. Observe the nail. What happened?
6. What can you say about the effect of heat on solids? Discuss your ideas with your group.

For Problem Solvers

Did you ever notice how hard it is to remove the ring from a jar of fruit? Try running hot water over the lid; then remove it. What do you think makes the difference?

Why are sidewalks made with joints every few feet? See what you can learn about expansion joints. See whether you can find expansion joints as you drive across overpasses or bridges on the highway. Why are these joints built into the bridge? Try to find expansion joints in large buildings.

If you know an automobile mechanic, ask him or her why wheel bearings are sometimes heated before they are installed on axles.

Teacher Information

Caution: This activity uses an open flame, necessitating close supervision.

As the candle heats the wire, the wire will expand. The nail, which was swinging freely above the surface, will drag. As it cools, the wire will contract and the nail will swing freely again. Other solids expand and contract similarly with temperature change.

If students have had experience with home-canned fruit, ask what they do when they cannot get the ring off the fruit jar. Most people learn, at some time, to run hot water over the ring, but many people do not know why the technique works. An excellent application for what has been learned in this activity would be to challenge students to explain why people run hot water over the ring. If they can explain this, they should be able to explain why the ring became so tight in the first place.

Integrating

Math, language arts

Science Process Skills

Observing, inferring, predicting, communicating, comparing and contrasting, using space-time relationships, formulating hypotheses, identifying and controlling variables

What Happens to Liquids as They Are Heated and Cooled?

Materials Needed

- Narrow-necked jars with one-hole rubber stoppers
- Balloon sticks (plastic tubes, available from craft or party-supply stores)
- Rubber bands, masking tape, or markers
- Bowls
- Cold water
- Food coloring

Procedure

1. Fill a jar completely with cold water.
2. Put several drops of food coloring into the water.
3. Insert the plastic tubing through the rubber stopper until it extends slightly out the bottom of the stopper.
4. Place the stopper in the jar. As you press the stopper into place, there should be no air space beneath the stopper, and water should be forced partway (not more than halfway) up the tube above the stopper. (See Figure 2.13–1.)
5. Mark the tube at the water level with a marker, or by putting a rubber band or tape around it.
6. Place the jar in the bowl, to catch any possible spills.
7. Place the bowl and jar in a window in direct sunlight.
8. Check the water level in the tube every few minutes for at least two hours.
9. What happened to the water level as the water warmed in the sunlight?
10. Remove the jar from the sunlight and put it in a cool place.
11. Again check the water level in the tube every few minutes.
12. What happened to the water level as the water cooled?
13. What can you say about the effect of temperature change on liquids?

Figure 2.13-1. Bottle Full of Water

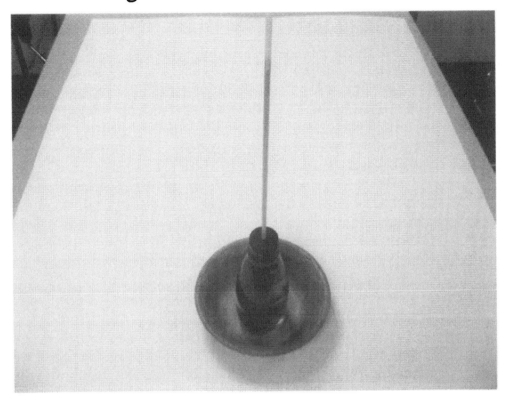

For Problem Solvers

The device you just made can become a thermometer. Place it on a table, next to the wall. Behind the tube, put a piece of paper that you can write on. Make a mark on the paper at the water level in the tube. Read the current temperature from a commercial thermometer and write the temperature beside the mark. As the temperature changes, record the temperatures on your paper. Then put the commercial thermometer aside and use your homemade thermometer to check the temperature each day. Do you think your thermometer will respond as rapidly to temperature changes as the commercial thermometer does? Why or why not?

If a motorist goes to the filling station on a hot day and fills the fuel tank clear to the brim, then parks the vehicle in the sun for two or three hours, sometimes the tank will overflow and spill fuel onto the ground. Think about the above activity and see whether you can explain why the tank overflows. The manager at your local filling station would probably be glad to discuss it with you if you have any questions or if you'd just like to find out if your explanation is correct. Any experienced truck driver could also discuss it with you.

Teacher Information

The ideal tube for this activity is a balloon stick, available at party supply outlets.

As the jar of water warms in the sunlight, the water will expand and the water level will rise in the tube, demonstrating that as the temperature of a liquid increases the liquid expands. As the water cools, it will contract, and the level of the water in the tube will drop.

If food coloring is available, have students add a few drops to the water. This makes the water level in the tube easier to see, and the change is more evident.

This device can become a thermometer if you have students attach a card to the tube and mark the card at different temperatures by taking temperature readings from a commercial thermometer. Having been calibrated, this thermometer will measure atmospheric temperature with considerable accuracy. It will be relatively slow to respond to temperature change because of the relatively large amount of liquid involved.

Water evaporation in the tube will eventually destroy the accuracy of it as a thermometer and it will need to be recalibrated.

Filling stations store gasoline in large tanks beneath the ground. Thus, the fuel is cool. If a motorist fills the tank, then parks the vehicle in the sun for a time, the fuel will expand from the heat and will sometimes overflow onto the ground.

Integrating

Math, social studies

Science Process Skills

Observing, inferring, classifying, measuring, predicting, communicating, comparing and contrasting, using space-time relationships, formulating hypotheses, identifying and controlling variables, researching

What Happens to Gases as They Are Heated and Cooled?

Materials Needed

- Narrow-necked jars with one-hole stoppers
- Balloon sticks (available from craft or party-supply stores)
- Water
- Bowls

Procedure

1. Put a small amount of water in the bottom of a jar.
2. Insert a plastic tube through the stopper.
3. Place a stopper in the jar. The lower end of the tube must be in the water. (See Figure 2.14–1.)

Figure 2.14–1. Bottle Nearly Empty

4. Notice the water level in the tube.

5. Place the jar in a window in direct sunlight. Put it in the bowl, to catch any spills.

6. Check the water level in the tube every three or four minutes for at least one-half hour.

7. What happened to the water level as the air warmed in the sunlight? Why?

8. Remove the jar from the sunlight and place it in a cool place.

9. Again check the water level in the tube every few minutes.

10. What happened to the water level as the air cooled? Why?

11. What can you say about the effect of temperature change on gases?

For Problem Solvers

Blow up a balloon and measure the circumference (distance around it) with a string. Mark the string to show the length required to reach around the balloon. Place the balloon over a heat vent or in front of a heater for a few minutes. Use the same string to measure the distance around the balloon again. Is there a difference? Can you explain why?

Learn what you can about hot-air balloons. Why do they rise into the air? Why do they come down again? Do balloon pilots usually fly their ships in the cool air of morning or in the heat of the afternoon? Why?

Teacher Information

Although both liquids and gases expand and contract as temperature changes, in this activity the change in water level in the tube is caused mostly by expansion and contraction of air within the jar. Air expands and contracts far more than do liquids with any given change in temperature. As the air in the jar warms in the sunlight, it will expand, forcing water up the tube and very likely spilling it out the top of the tube, demonstrating that as the temperature of a gas increases, the gas expands. As the air cools, it will contract and the level of the water in the tube will drop.

If food coloring is available, add a few drops to the water to make the water level in the tube more visible.

Integrating

Math

Science Process Skills

Observing, inferring, measuring, predicting, communicating, comparing and contrasting, using space-time relationships, formulating hypotheses, identifying and controlling variables, experimenting

What Other Type of Energy Accompanies Light from the Sun?

(Teacher-supervised activity)

Materials Needed

- Magnifying glasses
- One sheet of paper per small group
- Glass bowls

Procedure

1. Ask your teacher where you should do this experiment. You will need to be in bright sunlight with no wind.

2. Put the piece of paper in the bowl.

3. Hold the magnifying glass between the paper and the sun so a beam of light focuses on the paper, as shown in Figure 2.15–1.

4. Notice that as you move the magnifying glass closer toward and farther from the paper, the point of light changes in size. Notice also that it gets brighter as it gets smaller.

5. Adjust the distance between the paper and the magnifying glass to make the point of light very small and bright.

6. Pull the magnifying glass back about 1 cm (1/2 in.) and watch the paper.

7. Do you see anything happening to the paper? If so, what, and why do you think it is happening? What kind of energy is causing this to happen? Discuss your observations and ideas with your group.

Figure 2.15–1. Magnifying Glass on Sheet of Paper in Glass Bowl

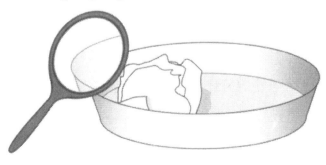

Teacher Information

As light rays from the sun are concentrated by a magnifying glass, so are the infrared, or heat, rays. The magnifying glass can concentrate bright sunlight to such a degree that it will scorch the paper or even possibly ignite it. As illustrated in Figure 2.15-2, the focal distance for heat (infrared) rays is slightly longer than the focal distance for light. For this reason, students are instructed to move the lens back slightly after the focal point of light is found. They will need to adjust the distance from lens to paper slightly to find the focal point for heat—the point at which the greatest possible concentration of heat is focused on the paper.

Integrating

Math, language arts

Science Process Skills

Observing, inferring, classifying, measuring, predicting, communicating, using space-time relationships, formulating hypotheses, identifying and controlling variables, experimenting

Figure 2.15-2. Focal Distance for Heat and Light

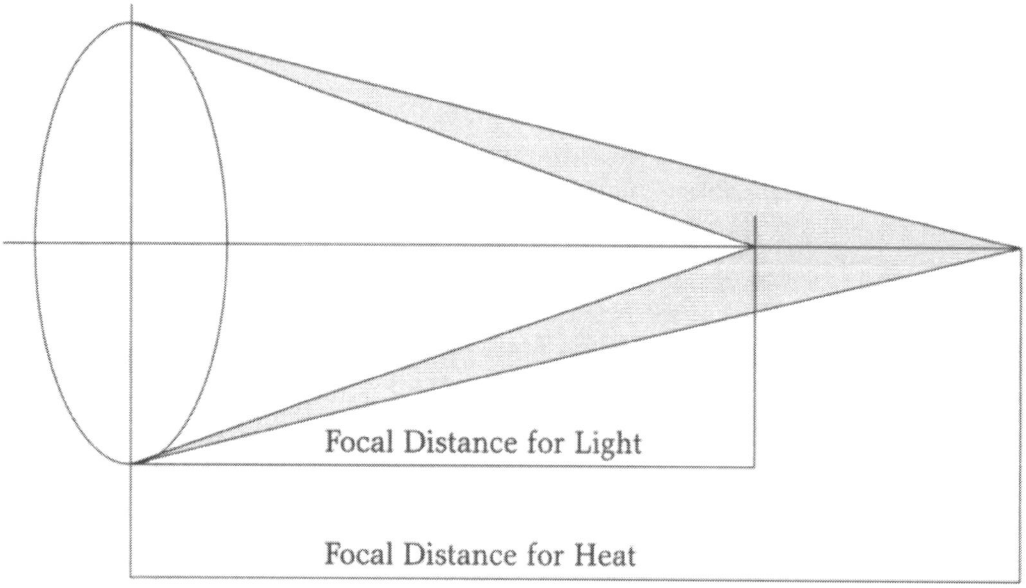

Focal Distance for Light

Focal Distance for Heat

How Can You Get the Most Heat Energy from the Sun?

Materials Needed

- Three identical jars per small group
- Paper and pencils
- Black paper
- Aluminum foil
- Tape
- Three thermometers per small group
- Sand

Procedure

1. Fill the three jars with sand.
2. Cover one jar with black paper, including the top, and tape the paper in place.
3. Cover the second jar with aluminum foil, including the top, and tape the foil in place.
4. Leave the third jar uncovered.
5. Record the temperature shown on the thermometers. Be sure all three indicate the same temperature.
6. Insert one thermometer into the sand in each jar. With the two covered jars, puncture a hole in the top covering and insert the thermometer through the hole.
7. Place all three jars in sunlight. All should receive the same direct sunlight.
8. Check and record the temperature of the three thermometers every 15 minutes for about two hours.
9. How do the temperatures compare? What can you say about the effect of a black surface and a shiny surface on absorption of energy from the sun?
10. Remove the jars from the sunlight and continue to record the temperatures of the three thermometers for two more hours.
11. How do the temperature changes compare? What can you say about the effect of a black surface and a shiny surface on heat loss?

For Problem Solvers

Here are more ways to compare the effect of color on heat absorption. Place a thermometer on a paper plate and lay a sheet of black paper over it. Prepare a second plate using a thermometer and white paper, then a third using aluminum foil. On a warm, sunny day, place all three plates in direct sunlight, being sure all papers are facing the sun at the same angle. Check and record the temperatures on the thermometers after one-half hour, then again after one hour.

If you live in a cold climate, place papers of various colors on a snow bank, in the sunlight. Use a black paper, a white paper, and a piece of aluminum foil of the same size. Use other colors also if you'd like to. Line them up so they all have direct sunlight and all papers are facing the sun at the same angle. After an hour, check the snow under the papers to find out how much has melted. Check them again after two hours.

Share your results with your class.

Teacher Information

The uncovered jar of sand will provide a control to help students observe the effect of both the black surface and the reflective surface. The temperature of the jar with the black surface will likely increase noticeably faster than that of the other two. The foil will reflect heat, and the temperature increase of the sand covered by it will be very slow.

Buildings in warm climates are typically covered with light-colored materials or light-colored paint. Astronauts wear reflective clothing to help protect them from the direct rays of the sun.

Integrating

Math

Science Process Skills

Observing, inferring, classifying, measuring, predicting, communicating, comparing and contrasting, using space-time relationships, formulating hypotheses, identifying and controlling variables, experimenting

How Does Color Affect Energy Absorbed from Light?

Materials Needed

- A copy of "Science Investigation Journaling Notes" for each student
- Two equally calibrated thermometers per small group
- Heat lamps (or high-wattage bulbs)
- One sheet of white paper and one sheet of black paper of the same thickness for each small group
- Paper and pencils

Procedure

1. Do the following as a Science Investigation. Obtain a blank copy of the "Science Investigation Journaling Notes." Write your name, the date, and your question at the top. Plan your investigation through item 5 of the form (Procedure) and have it approved by your teacher. Complete the Journaling Notes as you perform your investigation. When you are finished, share your project with your group and submit your Journaling Notes to your teacher if requested.

2. Prop the thermometers in an upright position about 20 cm (8 in.) apart, facing in the same direction.

3. Record the temperatures of both thermometers.

4. Place the sheet of white paper in front of one thermometer and the black paper in front of the other.

5. Shine the heat lamp at the sheets of paper in such a way that it faces both equally. It should be about 40 to 50 cm (16 to 20 in.) away from the papers. (See Figure 2.17–1.)

6. After the heat lamp has shone on the papers for about two minutes, check and record the temperatures of the two thermometers again.

7. Repeat for two more minutes and again record the temperatures.

8. Compare the changes in the first and last temperature readings of the two thermometers.

9. What can you say about the effect of color in this activity?

10. Complete your "Science Investigation Journaling Notes." Are you ready to explain how color affects energy absorption from light? Discuss it with your group.

Figure 2.17-1. Heat Lamp, Sheets of Paper, and Thermometers

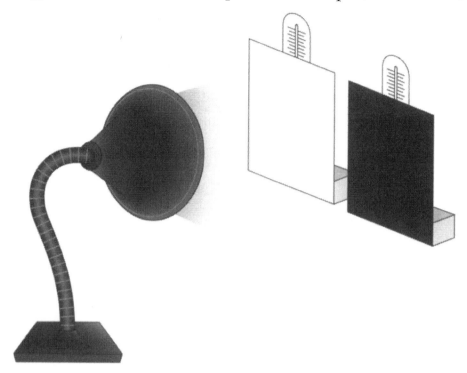

For Problem Solvers

In the above activity, you compared heat absorption from white paper and black paper. Find several different colors of paper. Use only paper that is the same except for color. Construction paper will work well. From what you have already learned, predict which colors will absorb the most heat from the light bulb and which will reflect the most heat. Arrange the papers in order, according to your predictions. Design an investigation to find out whether your predictions are accurate.

Were you right? Why is it important that you don't use construction paper for some colors, typing paper for some colors, and art paper for still other colors?

That raises another interesting question. Does one type of paper absorb more heat than another? To test that question, what will you do about color?

Teacher Information

Dark materials have a greater tendency to absorb heat than do lighter-colored materials. The thermometers behind the two sheets of paper will verify this tendency.

On a sunny day, direct sunlight could be used instead of the heat lamp. As an enrichment (or perhaps introductory) activity, invite students to go to the parking lot (on a warm day) and feel the surfaces of cars of various colors. Which colors tend to be warmest? This needs to be carefully supervised to avoid offending car owners.

Your "problem solvers" will enjoy the challenge of some real science while they test the variables of color and types of paper with respect to heat absorption.

Integrating

Math

Science Process Skills

Observing, inferring, classifying, measuring, predicting, communicating, comparing and contrasting, using space-time relationships, formulating hypotheses, identifying and controlling variables, experimenting

Where Does Our Energy Really Come From?

Materials Needed

- Paper (divided into two columns)
- Pencils

Procedure

1. Divide your paper in half with a vertical line.
2. Go for a walk around your home, your neighborhood, and around the school yard, looking for clues about solar energy.
3. In the left column of your paper, make a note of anything you see that involves solar energy in some way. Also include anything that you think has used solar energy in the past.
4. In the right-hand column, describe how you think solar energy is involved with each item on the left.
5. Share your observations with others who are doing this activity, and discuss your thoughts and ideas about it.

For Problem Solvers

Consider your observations of solar energy from this activity, along with all of the other energy-users you have previously observed. With this information, try to trace all of the energy sources back to the sun.

Go to the Internet, encyclopedia, and other sources and see what information you can find about the current state of solar energy development.

Teacher Information

Student observations might include such things as a solar collector on the roof of a house, a solar-powered calculator, or simply light and heat coming from the sun through windows. And since we're looking for clues of energy having been used in the past, as well as the present, what about the car parked along the street; the bricks that required energy to make, to transport, to set in place during the construction of the building; and so on.

Your problem solvers should consider all energy types and trace them to the sun. For example, plants get their energy from the sun; the plants could not grow without it. We all get our energy from plants and from animals that grew by eating the plants. The vehicles passing by are powered by gasoline, which comes from oil that was formed from organic matter of past ages. From the same oil comes the plastic bag we store food in and parts of the refrigerator that keeps the food fresh. Even wind energy results from the heating of air masses by the sun.

In doing this activity, students should conclude that almost all energy used on the earth can ultimately be traced to the sun.

Integrating

Language arts, social studies

Science Process Skills

Observing, inferring, math, communicating, researching

How Can You Cook an Apple in Your Own Solar Cooker?

Materials Needed

- Foam cups
- Clear plastic cups, sized to fit snugly inside the foam cups
- Three or four slices of pealed apple (about .5 cm [1/4 in.] thick) per small group
- Clear plastic wrap
- Clear plastic tape
- Black construction paper (8½ by 11 in.)
- Aluminum foil, about 75 cm by 75 cm (30 in. by 30 in.)

Procedure

1. Place the apple slices inside the clear cup.
2. Wrap the cup tightly with plastic wrap and secure with tape. (See Figure 2.19–1.)
3. Line the foam cup with black paper and tape the paper in place.
4. Place the clear cup inside the foam cup.
5. Wrap the foil around the cups in a funnel shape (shiny side in) that will reflect sunlight onto the top of the clear covering of the plastic cup. The top of the foil funnel should extend far above the cup.
6. Tape the foil in place if needed.
7. Place your solar cooker in the sunlight, positioned with the funnel opening toward the sun. You might need to reposition the cooker a time or two to keep the funnel facing directly into the sun, as the earth turns a bit on its axis.

Figure 2.19-1. Solar Cooker

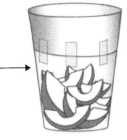

Apples in clear plastic cup, wrapped in plastic
kitchen wrap (secured with tape) →

Foam cup lined with
black construction paper →

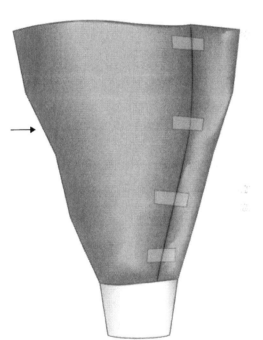

Funnel made of aluminum foil
(secured onto the foam cup with tape) →

For Problem Solvers

Think of ways to improve the cooker for a more efficient design. What variables might you change in your investigation? Size of the cups? Omit the foam cup? Is the plastic wrap needed? What about the size of the foil funnel? Is the funnel needed at all? Can you cook a hotdog? What else?

Go to the Internet and see what you can learn about current technology in the use of solar energy for cooking, especially in undeveloped countries.

Teacher Information

Caution: Most school systems do not allow food prepared in the classroom to be eaten by students, but if eating the food is allowed, try some cinnamon and sugar on the apple slices.

Here's a simple design for students to experiment with, and perhaps use as a springboard to investigate with design changes.

The plastic wrap provides an enclosure to take advantage of the greenhouse effect. The foam cup insulates the plastic cup to minimize heat loss.

You might consider kicking off some competition for the most efficient new design students can develop for a solar cooker.

Integrating

Reading, language arts, math, social studies

Science Process Skills

Observing, inferring, measuring, predicting, communicating, identifying and controlling variables, experimenting, researching

How Does Gravity Affect Heavy and Light Objects?

(Take home and do with family and friends.)

Materials Needed

- "Science Investigation Journaling Notes" for each student
- Large book
- Small book
- Wadded paper
- Pencil
- Eraser
- Paper clip
- Paper

Procedure

1. Do the following as a Science Investigation. Obtain a blank copy of the "Science Investigation Journaling Notes." Write your name, the date, and your question at the top. Plan your investigation through item 5 of the form (Procedure) and have it approved by your teacher. Complete the Journaling Notes as you perform your investigation. When you are finished, share your project with your group and submit your Journaling Notes to your teacher if requested.

2. Take the large book in one hand and the small book in the other. Hold the two books at exactly the same height.

3. Drop both books at the same time, but before you drop them, predict which one will fall faster. Have someone watch to see which book hits the floor first.

4. Repeat the book drop three times to be sure of your results.

5. Which book falls faster, the large one or the small one?

6. Compare the pencil and the paper in the same way. First predict which you think will fall faster.

7. Compare the various objects, two at a time. In each case, predict which will fall faster, then drop them together three times to test your prediction.

8. Of all the materials you tried, which falls fastest? Most slowly?

9. Explain how the force of gravity compares with objects that are large, small, heavy, and light, according to your findings. How do the falling speeds compare?

10. Compare the falling speed of the wadded paper with that of a flat sheet of paper dropped horizontally.

11. Compare the falling speeds of two flat sheets of paper, one dropped vertically and the other horizontally.

12. Compare the falling speed of the wad of paper with that of a flat sheet of paper dropped vertically.

13. Discuss your observations with your group.

14. Complete your "Science Investigation Journaling Notes." Are you ready to compare the effect of gravity on heavy and light objects? Discuss it with your group.

For Problem Solvers

Go to the Internet, the encyclopedia, and other resources and do some research about gravity. Can you find out what really causes gravity? How large does an object have to be in order for it to have a gravitational attraction for other things? How much do scientists know about gravity? Do they really know what it is?

Teacher Information

The force of gravity pulls all falling objects to the earth at the same rate, regardless of the size or weight of the object. Air resistance can slow the rate of fall, so the flat paper held in horizontal position will fall more slowly. Except for the factor of air resistance, however, the rate of fall is equal. A brick and a feather will fall at the same speed if placed in a vacuum chamber.

Scientists are still trying to figure out exactly what gravity is. They have learned a great deal about it. They know that all objects have a gravitational attraction for all other objects, although the force is too weak to really notice unless the objects are huge, as with planets and stars.

Integrating

Reading, math

Science Process Skills

Observing, inferring, predicting, communicating, comparing and contrasting, experimenting, researching

What Is Center of Gravity?

(Take home and do with family and friends.)

Materials Needed

- "Science Investigation Journaling Notes" for each student
- Meter stick (or yardstick)
- String
- Chair
- Various books

Procedure

1. Do the following as a Science Investigation. Obtain a blank copy of the "Science Investigation Journaling Notes." Write your name, the date, and your question at the top. Plan your investigation through item 5 of the form (Procedure) and have it approved by your teacher. Complete the Journaling Notes as you perform your investigation. When you are finished, share your project with your group and submit your Journaling Notes to your teacher if requested.

2. Balance the meter stick on the back of the chair. It will balance at its "center of gravity," which should be at or very near the 50-cm (18-in.) mark. The part of the chair where the meter stick rests is the *fulcrum*.

3. Get two identical books and tie a string around each one.

4. Make a loop in the other end of each string and slide the loops over opposite ends of the meter stick. Leaving the books supported at the ends of the meter stick, where is the center of gravity (where the fulcrum has to be to balance the books)?

5. Replace one of the books with a smaller book. With the books still suspended at the ends of the meter stick, where is the center of gravity?

6. Replace the other book with a larger one. Where is the center of gravity now?

7. What can you say about the center of gravity when a large object is balanced with a small object? Consider the teeter-totter as you explain your answer.

8. Complete your "Science Investigation Journaling Notes." Are you ready to explain what the center of gravity of an object is? Discuss it with your group.

For Problem Solvers

Do you know what a mobile is? Build one, and then explain why it is important to know about center of gravity when constructing mobiles. If you do not know what a mobile is, ask your teacher, a parent, or a friend to help you get started.

Teacher Information

This activity is closely related to the activities on first-class levers in the section "Simple Machines." Center of gravity is the balance point of an object. The center of gravity of spherical objects is at the center, assuming, of course, that the mass is equally distributed throughout the object.

Mobiles are fascinating to construct and provide excellent application of the concept of center of gravity. Encourage students to make a mobile, as suggested in the "For Problem Solvers" section.

Integrating

Math

Science Process Skills

Observing, inferring, identifying and controlling variables

Where Is Your Center of Gravity?

(Take home and do with family and friends.)

Materials Needed

- Pencil or other small object

Procedure

1. Put your pencil on the floor.
2. Standing near the pencil, pick it up without bending your legs or moving your feet.
3. Stand against the wall, with your heels touching the wall.
4. Drop your pencil on the floor near your feet.
5. Bend over and pick up your pencil without moving your feet or bending your legs. You must also not lean against anything or hold onto anything for support.
6. What happened? Why?
7. Repeat steps 1 and 2. Notice your movements as you pick up the pencil. Explain what happened in step 5 in terms of the effect of the center of gravity.

For Problem Solvers

Try this activity with family members and friends. Try replacing the pencil with a coin or with a dollar bill. Can you find anyone who can pick up the object without breaking the rules?

Teacher Information

Any time we are on our feet, whether we are walking, running, standing, or bending over, we are constantly adjusting to the center of gravity in order to remain "balanced." The body makes these adjustments so automatically that we don't think about them. Our feet, or some part of our body, must remain under our center of gravity, or we will fall.

The "For Problem Solvers" section invites students to try this activity with family members. Replacing the pencil with a dollar bill and offering it to the person who can pick it up without breaking the rules will increase interest and effort substantially. The money is as safe as if it were behind lock and key.

Integrating

Physical education, dance

Science Process Skills

Observing, inferring, communicating

How Can You Balance Several Nails on One Nail?

(Take home and do with family and friends.)

Materials Needed

- Wood base with nail hole in center
- Several flat-headed nails

Procedure

1. This is an activity you can use to challenge your family and friends. After they give up, you can show them how smart you are.
2. Stand one nail (we'll call it number 1) in the hole in the board.
3. Lay a second nail (number 2) on the table.
4. Place the remaining nails on the table with their heads over nail number 2, in alternating directions.
5. Lay the last nail on top of nail number 2, where the nail heads overlap.
6. Pick up nail number 2 and the last carefully by the ends. If you have set it up correctly, the criss-crossed nails will come, too.
7. Carefully balance the whole system on nail number 1. The entire setup is shown in Figure 2.23–1.

Figure 2.23–1. Balancing Nails

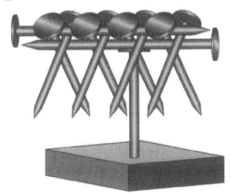

For Problem Solvers

What is the role of center of gravity with the nail-balancer? Why don't the overhanging nails tip the system over?

Did you ever see a tightrope walker? They usually carry a low-bending pole. Can you see a similarity with the nail-balancer? Do some research about tightrope walkers and learn what you can about the way they use center of gravity to their advantage.

Teacher Information

First, you might want to just give the materials to students with verbal instructions to stand one nail in the wood base and balance all of the other nails on the one that's standing. Let them struggle with it for a while. The instructions and illustration will help students when you're ready for them to have it.

Although nails are stretching out in two directions, the fourteen nails will balance on the first nail, because the center of gravity is at the center of the system of nails. The overhanging nails actually hold the system in balance instead of tipping it over, as one might expect at first glance. Tightrope walkers use low-bending poles to help them balance for the same reason.

This is a great activity for students to do at home. Encourage them to let their victims struggle; don't be in a hurry to show them how to do it.

Integrating

Math, language arts

Science Process Skills

Observing, inferring, communicating, researching

What Happens When You Burn a Candle at Both Ends?

(Teacher demonstration)

Materials Needed

- Candle
- Match
- Round toothpicks
- Water glasses

Procedure

1. Prepare a candle so the wick may be lighted at both ends.
2. Insert a round toothpick into each side of the candle and balance it on the water glasses as shown in Figure 2.24-1. It doesn't have to balance perfectly.

Figure 2.24-1. Candle Burning at Both Ends, Balanced Between Glasses

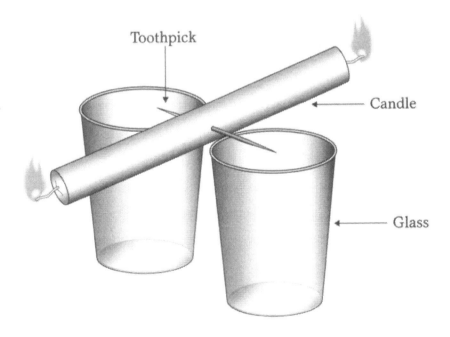

3. Have students predict what will happen if you light both ends of the candle.

4. Light both ends of the candle. Observe for several minutes. Ask what happened.

5. Have students think about it and explain what happened the best they can. Have them share their ideas with others in the group.

Teacher Information

Caution: Activities with fire or heat should be done only under close adult supervision or as a teacher demonstration.

When the candle is lighted at both ends, the end tilting downward will burn wax away more rapidly and become lighter. When it tilts up, the other end will be down, and it will burn wax away more rapidly. As this process continues to reverse, the candle will rock back and forth, often quite vigorously. The center of gravity continues to adjust, with the loss of weight from one end, then the other as the wax drips away.

Integrating

Math, language arts

Science Process Skills

Observing, inferring, predicting, communicating, formulating hypotheses

How Does Inertia Affect the Way Things Move?

(Teacher demonstration)

Materials Needed

- Two chairs
- Broom
- Four lengths of lightweight cotton string about 45 cm (1.5 ft.) each
- Two bricks, rocks, or other weights, about 1 kg (2 lbs.) each

Procedure

1. Lay the broom across the backs of two chairs (or other supports).
2. Tie two pieces of string to the broom handle, several inches apart.
3. Tie each brick to one of the strings hanging from the broom handle. The bricks should hang down several inches from the broom handle.
4. Tie one of the other two strings to each of the bricks. These strings should hang freely from the bricks. Each setup of brick and strings should be approximately as shown in Figure 2.25–1.
5. Hold tightly to one of the lower strings and pull down slowly but firmly until a string breaks.
6. Hold tightly to the other lower string and jerk quickly, breaking a string.
7. Ask students which string broke when you pulled slowly—the upper string or the lower string? Which one broke when you jerked? Have them explain.
8. Get some new string and repeat the activity to verify their results.
9. Ask why they think it happened that way. Do they see any effect of inertia? Have them discuss their ideas and observations.

Figure 2.25-1. Brick with Two Strings

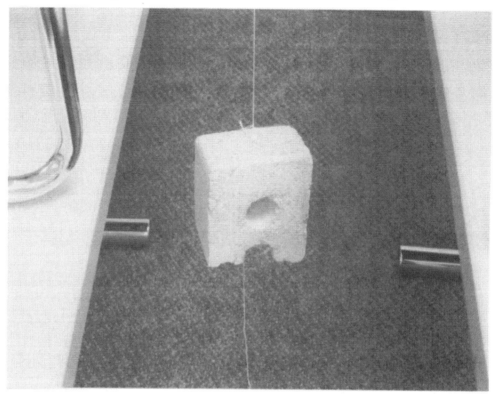

For Problem Solvers

Pick a spot (target) on the floor or the sidewalk. Walk to the target and stop with your feet exactly on it. Now run to the target and stop with your feet exactly on the spot. Next, run as fast as you can and try to stop with your feet exactly on the spot, without slowing down before you get there. What effect is inertia having on your effort to stop?

Think about seat belts. Why is it important to use them when riding in a motor vehicle? What does inertia have to do with the need for seat belts?

Why do the earth and the other planets remain in orbit around the sun, instead of being pulled into the sun by gravity? What do you think inertia has to do with this?

Teacher Information

Caution: This is recommended as a teacher demonstration because of the chance that a student could be injured if a hand is not moved quickly when the top string breaks and the brick falls.

As a lower string is pulled slowly, the force of the pull is equal on both the upper and lower strings. In addition, the upper string is supporting the weight of the brick (pull of gravity). Thus, the upper string will usually break if the two strings are identical.

Newton's first law of motion states that an object at rest tends to remain at rest, and an object in motion tends to remain in motion in the same direction and at the same speed unless an outside force acts upon it. This is sometimes referred to as the *law of inertia*. Inertia is the resistance to change in motion referred to in Newton's first law. The brick, in this case, is an object at rest, hanging from the string. As the lower string is given a quick jerk, the resistance, or inertia, of the brick protects the upper string from receiving the full impact of the sudden downward force. Thus, at that instant, greater force is applied to the lower string than to the upper string, and the lower string will usually break.

Integrating

Math, language arts

Science Process Skills

Observing, inferring, communicating, comparing and contrasting

What Is Centrifugal Force?

Materials Needed

- Small tubes, about 10 cm (4 in.) long
- String, about 1 m (1 yd.) long
- Two pencil erasers (or other small weights) per small group

Procedure

1. Thread the string through the tube.
2. Tie one pencil eraser to each end of the string.
3. Hold the tube upright and move it around in a circular motion so the top weight swings around and around, as shown in Figure 2.26–1.
4. Swing the weight around faster. Do not swing it near anyone. The tendency of the upper weight to move outward when rotating is commonly called centrifugal force.
5. Change the speed of rotation, faster and slower, and observe the lower weight.
6. What happens to the lower weight as you increase and decrease the speed of rotation? What can you say about the speed of rotation and its effect on centrifugal force? Discuss your ideas with your group.

Teacher Information

The tube used in this activity could be a cardboard tube. It could even be the barrel of a ballpoint pen or a short piece of plastic pipe. If a small, sturdy tube is not available, use a large wooden bead out of someone's toy box. The bead needs to be large enough to hold firmly in the hand without interfering with the movement of the string passing through it.

Centrifugal force is the force that tends to impel an object outward from a center of rotation. Newton's first law of motion states that an object at rest tends to remain at rest and an object in motion tends to remain in motion at the same speed and in the same direction unless it is acted upon by an outside force. The tendency of the object to continue moving in a straight line and at the same speed is called inertia. If the object is held back by another force, it cannot do that; thus the circular motion. The force that holds it back is called centripetal force, defined as

Figure 2.26-1. Tube, String, and Erasers, One Swinging

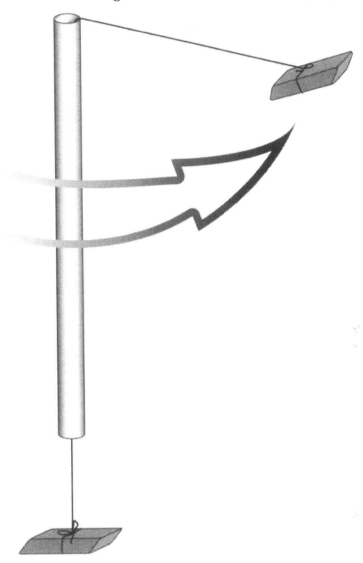

the force that tends to pull an object toward the center of rotation. Thus, what we call centrifugal force is really an interaction between inertia and centripetal force.

Integrating
Math, language arts

Science Process Skills
Observing, inferring, comparing and contrasting

Can You Solve This Energy Word Search?

Try to find the following Energy terms in the grid below. They could appear in horizontal (left to right), vertical (up or down), or diagonal (upward or downward) position.

kinetic	potential	energy
horsepower	pounds	inertia
weight	force	sun
wind	heat	sunlight
light	molecules	expands
gravity	source	

```
H   F   Q   A   E   X   P   A   N   D   S   D
L   O   G   P   O   I   O   W   I   N   D   F
K   R   R   T   K   H   W   K   N   J   H   G
K   C   A   S   S   K   I   N   E   T   I   C
T   E   V   U   E   R   A   R   Y   T   T   T
H   E   I   N   L   P   S   E   T   R   H   Y
G   P   T   Z   U   X   O   C   I   V   G   T
I   O   Y   N   C   B   U   W   A   R   I   H
L   U   P   M   E   O   R   I   E   U   E   G
N   N   E   R   L   T   C   N   Y   R   W   I
U   D   W   P   O   T   E   N   T   I   A   L
S   S   Q   C   M   V   B   N   M   K   J   H
```

Can You Create a New Energy Word Search of Your Own?

Write your Energy words in the grid below. Arrange them in the grid so they appear in horizontal (left to right), vertical (up or down), or diagonal (upward or downward) position. Fill in the blank boxes with other letters. Trade your Word Search with someone else who has created one of his or her own, and see whether you can solve the new puzzle.

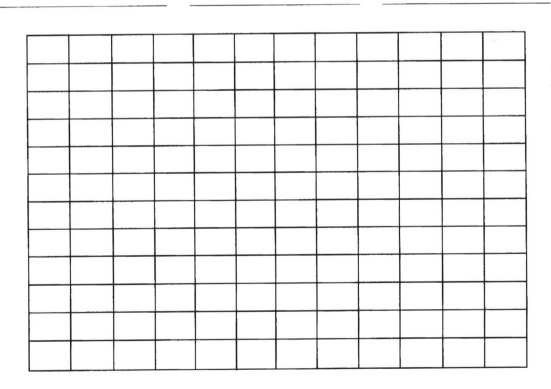

Answer Key for
Energy Word Search

```
H  F  Q  A  E  X  P  A  N  D  S  D
L  O  G  P  O  I  O  W  I  N  D  F
K  R  R  T  K  H  W  K  N  J  H  G
K  C  A  S  K  I  N  E  T  I  C  C
T  E  V  U  E  U  R  A  R  Y  T  T
H  E  I  N  L  P  S  E  T  R  H  Y
G  P  T  Z  U  X  O  C  I  V  G  T
I  O  Y  N  C  B  U  W  A  R  I  H
L  U  P  M  E  O  R  I  E  U  E  G
N  N  E  R  L  T  C  N  Y  R  W  I
U  D  W  P  O  T  E  N  T  I  A  L
S  S  Q  C  M  V  B  N  M  K  J  H
```

Do You Recall?

Section Two: Energy

1. Describe one example of kinetic energy.

2. Describe one example of potential energy.

3. Describe an act that involves doing five foot-pounds of work.

4. Describe wind energy; what is its source and how does it happen?

5. What other type of energy comes with light energy from the sun?

Energy

Do You Recall? *(Cont'd.)*

6. How does sound travel?

7. How can magnetism do work?

8. Why does a nail get hot as it is pounded into a board?

9. Why do substances expand as they get warmer?

10. If dark-colored objects and light-colored objects are exposed to sunlight, which colors are more inclined to absorb heat?

11. Think of a task you do almost every day. Trace the energy, step-by-step, all the way back to the sun.

Do You Recall? *(Cont'd.)*

12. Do heavy objects fall faster than light objects due to gravity, or do light objects fall faster? What about large objects compared to small objects?

13. Think about center of gravity. If you stand with both feet together and lean over, what determines when you fall?

14. Consider a situation in which you throw a ball straight in front of you. If neither gravity nor air existed, and no objects were in the way of the ball, how far would the ball travel before it stopped?

Answer Key for Do You Recall?

Section Two: Energy

Answer	Related Activities
1. Answers will vary.	2.3
2. Answers will vary.	2.3
3. Answers will vary.	2.4
4. Answers will vary, but it goes right back to the sun.	2.5
5. Heat energy	2.5
6. Vibrating molecules bump into each other.	2.6
7. Any time something is caused to move, work is done. Therefore, when a magnet attracts an object and causes the object to move, the magnet does work.	2.7
8. Due to friction, energy from the moving hammer changes to heat energy.	2.10
9. Molecules move more rapidly and with greater energy.	2.11–2.14
10. Dark colors	2.17
11. Answers will vary.	2.18
12. All objects fall at the same rate due to gravity, unless the object is affected by air resistance.	2.20
13. When your center of gravity is no longer over your feet	2.21–2.24
14. The ball would never stop.	2.25

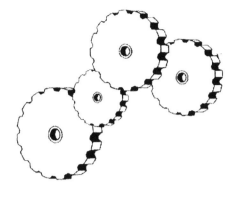

Light

To the Teacher

Like many other scientific phenomena, light is so common that we take it for granted. Yet without it we could not live. Plants use light from the sun to produce oxygen, which is vital to all animal life, including humans. Without plants, we would have no food. Light from the sun also heats the earth, and without heat there could be no life at all.

The question of what light really is has evaded scientists for centuries. Yet it is as fascinating as it is elusive and continues to be the object of many studies. We know a great deal about light because of these studies. For instance, we know that light is a form of energy that travels freely through space. We also know that, in addition to the sources of *natural* light (the sun and the stars), light can be created in various ways. When light comes from sources that people control, it is called *artificial* light. We use artificial light every day in the form of fluorescent lights and incandescent lights. The laser produces a form of light that has found widespread use in industry, medicine, and communications.

Activities included in this section encourage investigation into some of the ways in which light behaves. As students participate in these activities, you should encourage them to ponder the relationship of this topic to the study of the eyes and to art.

The scope of the activities in this section is limited to a few very basic concepts about light. Students investigate shadows, color, reflection, and refraction, and they are introduced to prisms and lenses. Many of these concrete activities are easily adaptable for children in the early grades. For the student whose interests extend beyond these basic investigations, many resources are available, including trade books, the Internet, encyclopedias, science reference books, and suppliers of scientific equipment.

The following activities are designed as discovery activities that students can usually perform quite independently. You are encouraged to provide students (usually in small groups) with the materials listed and a copy of the activity from the beginning through the "Procedure." The section titled "Teacher Information" is not intended for student use, but rather to assist you with discussion following the hands-on activity, as students share their observations. Discussion of conceptual information prior to completing the hands-on activity can interfere with the discovery process.

Suggestion with a caution: With several of the activities in this section that involve the use of the flashlight, the laser pointer is even more effective because of the well-defined beam of light that it produces. Laser pointers are in common use today and quite available in most schools. If the laser pointer is used, however, extreme caution must be taken to see that it is not pointed at the eyes.

Regarding the Early Grades

With verbal instructions and slight modifications, many of these activities can be used with kindergarten, first-grade, and second-grade students. In some activities, steps that involve procedures that go beyond the level of the child can simply be omitted and yet offer the child an experience that plants the seed for a concept that will germinate and grow later on.

Teachers of the early grades will probably choose to bypass many of the "For Problem Solvers" sections. That's okay. These sections are provided for those who are especially motivated and want to go beyond the investigation provided by the activity outlined. Use the outlined activities, and enjoy worthwhile learning experiences together with your young students. Also consider, however, that many of the "For Problem Solvers" sections can be used appropriately with young children as group activities or as demonstrations, still giving students the advantage of an exposure to the experience and laying groundwork for connections that will be made later on.

Correlation with National Standards

The following elements of the National Standards are reflected in the activities of this section.

K–4   Content Standard A: Science as Inquiry

As a result of activities in grades K–4, all students should develop:

1. Abilities necessary to do scientific inquiry
2. Understanding about scientific inquiry

K–4   Content Standard B: Physical Science

As a result of activities in grades K–4, all students should develop understanding of

1. Properties of objects and materials
2. Position and motion of objects
3. Light, heat, electricity, and magnetism

5–8   Content Standard A: Science as Inquiry

As a result of activities in grades 5–8, all students should develop:

1. Abilities necessary to do scientific inquiry
2. Understanding about scientific inquiry

5–8   Content Standard B: Physical Science

As a result of activities in grades 5–8, all students should develop understanding of

1. Properties and changes of properties in matter
2. Motions and forces
3. Transfer of energy

What Can You Make with a Shadow?

(Take home and do with family and friends.)

Materials Needed

- Projector or flashlight
- Wall or screen

Procedure

1. Using a blank wall or a piece of paper or cardboard as a screen, see whether you can make the animal shapes shown in Figure 3.1–1 with shadows.

Figure 3.1–1. Shadow Pictures

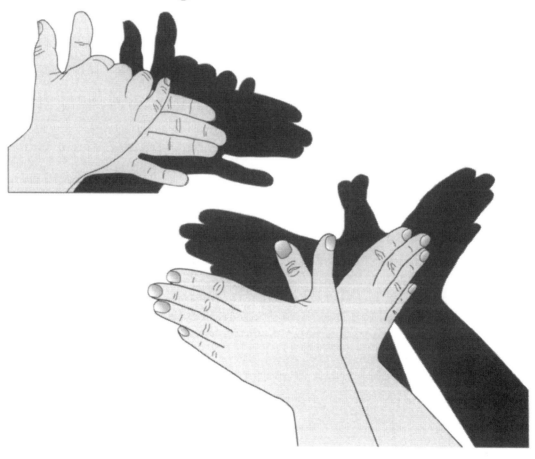

2. Create some other shapes of your own.

3. What happens to the shadow figures as you move your hands closer to or farther from the light?

4. Where is the darkest part of the shadow? Explain why it is darker there.

For Problem Solvers

Find a white bed sheet, a projector, and some way to hang the sheet up so you can shine light on it from the projector. Get some friends to help you create a shadow play from behind the sheet. Some members of your group can create the images while the rest are watching them from the front of the sheet.

Here's a challenge for you. Have one member of your group go behind the sheet and make a shadow on the sheet with his or her body. Then have a member of the group in front of the screen try to make the same body shape by looking at the shadow. Someone else can move the sheet and check to see whether the two body shapes are about the same.

Now set up an overhead projector, shining on a wall or screen. Find some three-dimensional geometric shapes, such as a cube, a cylinder, a rod, or a sphere. Stand a book or other visual barrier at the back of the projector plate so others can't see the item(s) you place on the projector plate. Make shadows on the wall (or screen) with these by placing them on the projector plate, while others try to guess which objects are being used to make the shadows. Try the same thing with any other object, such as a pencil, a thumbtack, an eraser, or whatever you can find. Turn the projector light off while placing the items on the plate.

Teacher Information

This activity is intended mostly for enjoyment and creativity. In reference to step 4, however, the students should notice that the shadow is darkest toward the middle. The outer edges of the figure are not shaded from the entire light source, as illustrated in Figure 3.1–2, and are therefore not as dark as the portion that is completely shaded.

The last part of "For Problem Solvers" works very nicely as a group activity using the overhead projector. Stand a book or other visual barrier at the back of the projector plate so students can't see the items being placed on the projector plate. Turn the projector light off while placing the item on the plate, then turn the light on and project the shadow on the screen.

As another group activity, divide the class into two groups. Have Group A stand behind the bed sheet. Have one person at a time from Group A stand in the light behind the screen and Group B try to guess who it is, based on the appearance of the shadow on the screen.

Figure 3.1-2. Light Creating a Shadow That
Is Darker in the Middle and Lighter on the Edges

Integrating

Language arts, art

Science Process Skills

Observing, inferring, comparing and contrasting

Why Does Your Shadow Change in Size and Shape?

Materials Needed

- Sheets of butcher paper or newsprint (one per student)
- Crayons or markers
- Partners
- Sunny day

Procedure

1. Lay a piece of paper on the sidewalk.
2. Stand on one end of the paper, so your entire shadow is on the paper.
3. Have your partner outline your shadow with a crayon and record the time of day.
4. Find a way to remember exactly where your paper was at the time your shadow was drawn so you can bring the paper back later and put it in exactly the same place.
5. Lie down on your piece of paper and see whether your body fits the outline of your shadow. If not, try to decide why not. Talk about it with your partner.
6. Trade places and outline your partner's shadow on a different piece of paper.
7. About an hour later, bring your piece of paper back and put it in exactly the same place that it was when your shadow was drawn.
8. Stand the same way you were when your shadow was drawn.
9. Does your new shadow fit your old one? Move around and try to make it fit. Has something changed? What is it? Why? Talk about your ideas with the others.

For Problem Solvers

With another color of crayon, draw your shadow the way you think it will look two hours later. Come back and check it out.

At mid-morning on a different day, have your partner draw your shadow again. Do it at the same time of day that you did before. When the drawing is finished, predict what your shadow will look like in the middle of the afternoon. Be sure that, when you come back to check it out, you put your paper in the exact same place and stand exactly as you did when the shadow was drawn.

Go to your Earth globe and select a spot on the globe that is at a different latitude from where you are. Then draw your shadow as you think it would be right now if you were there. Try the same thing again, but using your latitude with a different longitude.

What variables are you using in this activity?

Do some research and learn what you can about shadow clocks and sundials. Can you use your shadow as a clock? Talk about it with your group and figure out a way to make a shadow clock with your shadow.

Which do you think would make the best shadow clock, your shadow or the shadow of the flagpole? Why? See whether you can get permission to draw on the pavement with chalk and make lines for a shadow clock around the flagpole. Be sure to plan carefully and clear your plan with your teacher before you make any lines on the playground.

Teacher Information

Mid to late morning is a good time to do this activity. Each person will need a piece of butcher paper that is a bit longer than the person is tall.

In this activity, students learn that shadows aren't always the same size and shape as the objects that cast the shadows. They will also discover that shadows change in size, shape, and direction as the day goes on and the sun moves across the sky.

Students who do the extension for problem solvers will have fun with the predictions. Notice whether they draw their predicted afternoon shadows in the opposite direction from their feet as their morning shadows were. They will also have an interesting challenge with the task of mentally placing themselves at a different latitude or longitude and drawing the shadows they would have if they were at that location.

Your problem solvers will also learn something about how shadows can be used to indicate time of day. If they decide to try making a shadow clock on the playground, you need to insist that they plan carefully and make it neatly so it will be accurate and so that others will use it. If the playground is not paved, students can draw their lines for the shadow clock in the soil.

Integrating

Math, reading, language arts, social studies, art

Science Process Skills

Observing, inferring, measuring, predicting, communicating, comparing and contrasting, using space-time relationships, formulating hypotheses, identifying and controlling variables, researching

150

What Path Does Light Follow?

Materials Needed

- Four 5- by 8-inch cards per small group
- Clay
- Flashlight or projector (or laser pointer)
- Rulers
- Paper punch

Procedure

1. Punch a small hole in the center of each card.
2. Stand each card in a small ball of clay.
3. Space the cards about 30 cm (1 ft.) apart, with the center holes lined up.
4. Have someone hold the flashlight so that it shines into the hole at one of the end cards while you look through the hole of the card at the other end. (See Figure 3.3-1.)
5. What do you see?
6. Now move one of the cards about one inch to the side and repeat step 4.
7. What happened? What does this tell you about the path light travels?

Figure 3.3-1. Light Shining Through Cards

Teacher Information

This activity will help children realize that light travels in a straight path. Unless the cards are positioned with the holes in a straight line, the light does not pass through the holes in the cards.

Caution: The laser pointer is more effective with this activity than is the flashlight or projector because of the well-defined beam of light that it produces. If the laser pointer is used, however, extreme caution must be taken to see that it is not pointed at the eyes.

Integrating

Math, language arts

Science Process Skills

Observing, inferring

How Does Light Energy Travel?

Materials Needed

- Jump ropes (a fairly long rope will work better than a short one)
- Partners

Procedure

1. Have a friend hold one end of the rope while you hold the other end.
2. Hold the rope about waist high and stretch it out quite tight.
3. Give your end a quick, short flip and observe the movements of the rope.
4. Ask your friend whether he or she felt anything at the other end of the rope.
5. Now ask your friend to give the rope a flip as you hold your end steady.
6. What did you feel? What did you see?
7. As one of you flipped the rope, did the rope travel, or just the energy?

Teacher Information

Students should conclude that, although the rope did not go anywhere, energy moved along the rope. There are many forms of wave energy; light is one of them.

Wave energy can be observed by dropping a small object into a pond of water and watching the waves move across the pond. Notice that neither leaves, nor ducks, nor any other floating objects move along with the wave, but instead they move up and down as the wave energy travels across the pool. The water does not travel as the energy is transferred. This is a characteristic of wave energy. Light is one form of wave energy.

Wave energy travels similarly when rock layers in the earth's crust shift against each other. Energy, in the form of an earthquake, can be felt from many miles away, yet nothing visible traveled that distance.

Integrating

Math, language arts

Science Process Skills

Observing, inferring

What Makes Things Transparent, Translucent, or Opaque?

Materials Needed

- Variety of objects that are transparent, translucent, or opaque
- Masking tape
- Paper

Procedure

1. Sort the objects into groups a, b, and c as follows:
   a. If you can look through the object and see other things clearly.
   b. If light comes through the object, but you can't see other things clearly through it.
   c. If light does not come through the object at all.

2. Using the masking tape, label the groups as follows: a = transparent, b = translucent, and c = opaque.

3. Place one piece of paper with each group of objects, and write the label for the group (transparent, translucent, opaque) at the top of the paper.

4. Think about your living room window, your bathroom window (or another bathroom window that you have seen), and your bathroom door. Write each of these on the paper according to where it would fit best.

5. Find things around the room or around the building that would fit each of these categories, and add them to your lists.

6. Think of still more things that you know that fit into each category and add them to your lists.

 Hands-On Physical Science Activities

7. Share your lists with others, and see whether you can find items on any of the lists that should be on a different list. Do you see anything that might work on two different lists?

8. Prepare a fourth list of items that have similar purposes but are sometimes made of a transparent material and sometimes are translucent or opaque. Discuss your ideas with one another.

Teacher Information

While this is basically a vocabulary exercise, it will build a concept that is important as students study the topic of light. It will also sharpen their observation skills and provide reason to practice being precise in expressing themselves clearly as they compare and contrast the different materials they find.

Encourage students (especially in the early grades) to continue to notice materials around them that are transparent, translucent, or opaque and to share their observations with each other. Have them continue to add to the lists. Younger children might be easily motivated to continue the exercise at home. The fourth list might be intriguing even for older children.

Integrating

Math, language arts

Science Process Skills

Observing, classifying, comparing and contrasting

What Is the Difference Between Reflected Light and Source Light?

Materials Needed

- Mirrors

Procedure

1. Look briefly at the light in your room.
2. Now hold the mirror so that you can see the room lights in the mirror.
3. Which of these is the "source light" and which is "reflected light"?
4. If you were to cover the source light, what would happen to the reflected light?
5. If you cover the reflected light, what happens to the source light?
6. Light from your desk enables you to see the desk. Is it source light or reflected light?
7. Does the sun give off source light or reflected light? What about the moon? Share and compare your ideas.

Teacher Information

We receive light by two means: from sources that produce light and from objects that reflect light. There are relatively few sources of direct light (source light); everything else we see gives off reflected light. Light sources include light bulbs, fluorescent light tubes, burning matches and other fire, and the sun. Other objects can be seen only when there is light to be reflected from one or more light sources.

Note: Certain biological organisms, including fireflies, produce light chemically. These are light sources; they produce their own light.

Integrating

Language arts

Science Process Skills

Observing, inferring, comparing and contrasting

What Is Unique About Reflected Light?

(Teacher demonstration)

Materials Needed

- Projector or flashlight (or laser pointer)
- Two chalky chalkboard erasers
- Mirror
- Darkened room

Procedure

1. Arrange the projector and the mirror so that the light from the projector can be focused on the mirror.
2. With the room darkened, shine the projector toward the mirror at an angle.
3. Clap the chalkboard erasers together lightly in the beam of light and have students notice the angle of the light beam as it approaches and as it leaves the mirror. How do they compare?

Setup for Activity 3.7

4. Change the position of the projector so the light from the projector shines toward the mirror from different angles. Use chalk dust as needed to keep the light beam visible.

5. Each time you change the position of the projector, have students compare the angle of the light beam approaching the mirror to the angle of the reflected light beam.

Teacher Information

When light is reflected from a mirror, it is always reflected at the same angle as the angle of the light from the source to the mirror. In other words, the angle of reflection is always equal to the angle of incidence (approaching angle).

Caution: The laser pointer is more effective with this activity than is the flashlight or projector because of the well-defined beam of light that it produces. If the laser pointer is used, however, extreme caution must be taken to see that it is not pointed at the eyes.

Integrating

Language arts

Science Process Skills

Observing, inferring, comparing and contrasting, formulating hypotheses, identifying and controlling variables, experimenting

How Is Light Reflection Like the Bounce of a Ball?

Materials Needed

- Copies of the "Science Investigation Journaling Notes" for this activity for all students
- Rubber balls
- Darkened room
- Mirrors
- Flashlight or projector (or laser pointer)
- Partners

Procedure

1. As you complete this activity, you will keep a record of what you do, just as scientists do. Obtain a copy of the form "Science Investigation Journaling Notes" and write the information that is called for, including your name and the date.

2. For this activity you will learn about light reflection. For item 1, the question is provided for you on the form.

3. Item 2 asks for what you already know about the question. If you have some ideas about how light reflects, write your ideas.

4. For item 3, write a statement of how light reflection is similar to a bouncing ball, according to what you already know, and that will be your hypothesis.

5. Now continue with the following instructions. Complete your Journaling Notes as you go. Steps 6 through 10 below will help you with the information you need to write on the form for items 4 and 5.

6. Bounce the ball from the floor as straight as you can.

7. Now bounce the ball to your partner.

8. Try bouncing the ball at different angles. Notice the angle of the ball's path as it approaches the floor and compare it with the angle of its path as it leaves the floor. How do they compare? (See Figure 3.8–1.)

Figure 3.8-1. Bouncing Ball

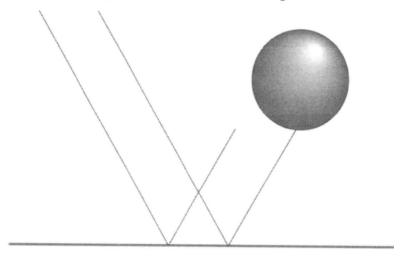

Figure 3.8-2. Bouncing Ball and Reflecting Light

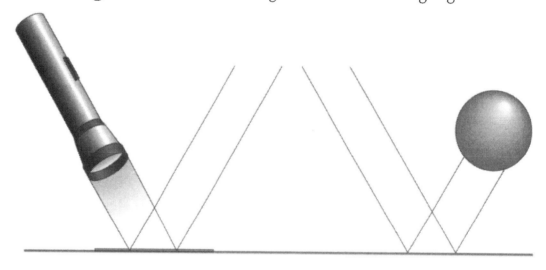

9. Now place the mirror on the floor and "bounce" the light from the flashlight off the mirror, first shining the light straight down at the mirror, then at different angles.

10. Compare the angles of light approaching and leaving the mirror. How do they compare with the angles of the bouncing ball approaching and leaving the floor? (See Figure 3.8-2.)

11. Complete your "Science Investigation Journaling Notes." Are you ready to explain how light reflection is like the bounce of a ball? Discuss it with your group.

For Problem Solvers

Does a "crazy ball" follow the same path when it bounces as light does when it is reflected? Try it.

Teacher Information

The purpose of the activity of bouncing the ball is to provide a familiar model. The student should notice that the angle of incidence (angle of the ball's path as it approaches the floor) equals the angle of reflection (angle of the ball's path as it leaves the floor). To avoid the curve in the ball's path created by the force of gravity, put a table next to a wall and roll the ball across the table to the wall.

Caution: The laser pointer is more effective with this activity than is the flashlight or projector because of the well-defined beam of light that it produces. If the laser pointer is used, however, extreme caution must be taken to see that it is not pointed at the eyes.

Integrating

Math

Science Process Skills

Observing, inferring, measuring, predicting, comparing and contrasting, formulating hypotheses, identifying and controlling variables

Science Investigation
Journaling Notes for Activity 3.8

1. Question: *How is light reflection like the bounce of a ball?*

2. What we already know:

_____

_____

3. Hypothesis:

_____

_____

4. Materials needed:

_____

_____

5. Procedure:

_____

_____

6. Observations/New information:

_____

_____

7. Conclusion:

_____

_____

How Many Images Can You See?

(Teacher-supervised activity)

Materials Needed

- Two mirrors per small group, one with a peephole in the center
- Small objects

Procedure

1. Hold the two mirrors a few centimeters (inches) apart with the reflecting surfaces facing each other, as in Figure 3.9-1.

Figure 3.9-1. Mirrors Facing Each Other, One with Peephole

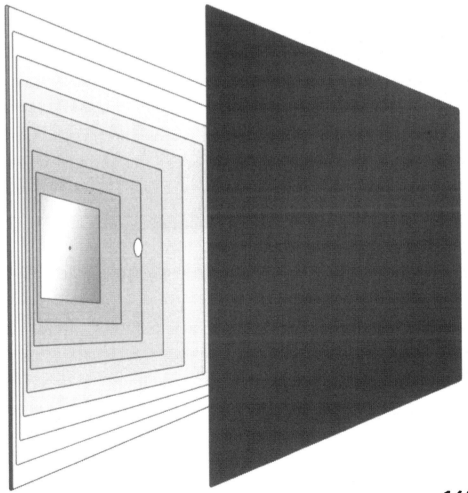

Figure 3.9-2. Small Object Held Between the Mirrors

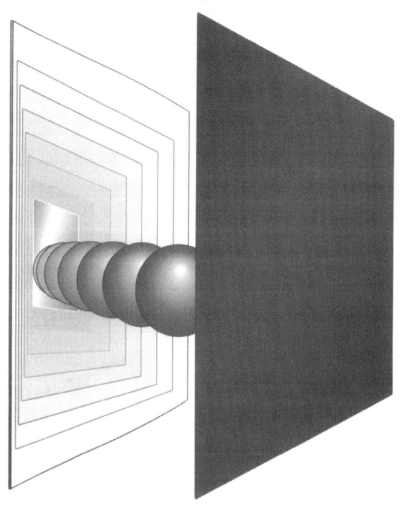

2. Hold a small object between the mirrors and look at it through the peephole, as shown in Figure 3.9-2.

3. What do you see?

4. Hold the object in different positions and tilt the mirrors at different angles.

5. Explain what you see and why you think it happens.

Hands-On Physical Science Activities

Teacher Information

To prepare the peephole mirrors, use a knife to scrape away the silvering from the back to form a small peephole about a centimeter (half-inch) in diameter right in the center of the mirror.

As light is reflected from one mirror to the other, an infinite number of images can be seen if the mirrors are kept parallel to each other. As the mirrors are held at a slant with respect to each other, fewer images will be seen because the slant brings the image closer to the top of the mirror with each reflection.

Integrating

Math, language arts

Science Process Skills

Observing, inferring, comparing and contrasting

How Well Can You Control the Reflection of Light?

Materials Needed

- Projector or flashlight
- Several partners
- One mirror for each person
- Darkened room

Procedure

1. Arrange the people with mirrors in a pattern such that light can be reflected from one to the other. (See Figure 3.10–1.)

2. From what you know about the reflected angles of light, see whether the group can direct the light from the projector to one mirror and have it reflected from the first mirror to a second mirror and from the second to a third.

Figure 3.10-1. Projector and Series of Mirrors

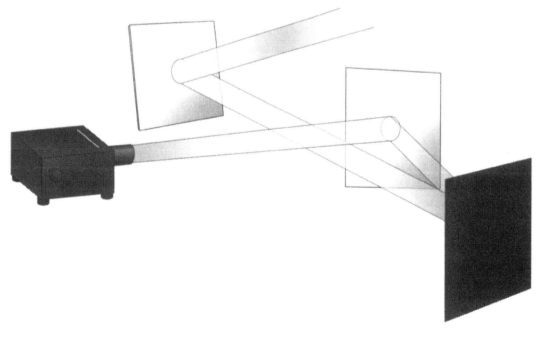

3. Determine who will reflect the light to whom in order to reflect the projector light all around the group.

4. Pick a spot (target) on the wall opposite the last person and light up the target with reflected light. Be sure the light from the projector reflects from all mirrors before lighting up the target.

5. With the light reflecting from all mirrors, compare the angle of reflection (the light leaving the mirror) with the angle of incidence (the light approaching the mirror) for each mirror. If necessary, have someone stand in the middle and clap two chalky chalkboard erasers together lightly to make the beams more visible.

6. How does the angle of reflection compare with the angle of incidence?

For Problem Solvers

With three friends, each of you with a mirror, play a game of "Reflection Relay." Select a target and position yourselves to reflect the light from a projector through all four mirrors and hit the target. Time yourselves and see whether you can improve your time with each new target you select. Challenge another group, and have a contest. Perhaps the entire class would like to get involved.

Teacher Information

Students will enjoy the challenge of reflecting the light from mirror to mirror in various patterns. They should notice that the angle of reflection and the angle of incidence are always equal.

Students who choose to try the reflected relay suggested in "For Problem Solvers" will acquire new insights about light and reflections. As they practice, they will learn that they need to stay near the light source and near one another, because the light spreads out and gets dimmer with distance. They will also learn to position themselves such that they are reflecting the light to one another as directly as possible, again to maximize brightness.

The use of laser pointers makes this activity less of a challenge because of its well-defined beam of light. If the laser pointer is used, there is value in using flashlights in addition, because flashlights are more commonly used in daily life. *Caution:* Be very cautious to see that laser pointers are never pointed into anyone's eyes.

Integrating

Math

Science Process Skills

Observing, measuring, predicting, communicating, comparing and contrasting, identifying and controlling variables, experimenting

Light

How Does a Periscope Work?

(Teacher-supervised activity)

Materials Needed

- Two one-quart milk cartons per small group
- Two mirrors (same width as milk carton) for small group
- Knife or scissors
- Tape

Procedure

1. Cut the tops off both milk cartons.
2. Cut an opening about 5 cm (2 in.) in diameter in one side of each carton, near the bottom.
3. Tape a mirror in the bottom of each carton, facing the opening at a 45-degree angle.
4. Tape the cartons together at the open ends to make a long tube, as shown in Figure 3.11–1.
5. Look into the mirror at one end of your periscope. What do you see?
6. Can you put your periscope together in such a way that you can look behind you? To your right?

For Problem Solvers

Design your own periscope. What can you do to make it longer? Can you make it so that you can turn it around 360 degrees, so you can see in any direction without turning yourself around? Do you need more mirrors in your design? Can you find round tubing, or something else that will work better than milk cartons?

Teacher Information

The periscope activity is likely to attract a lot of interest and could be used as an enrichment activity. It is an application of the concept of angle of reflection students learn about in other activities of this unit. Students might enjoy expanding their periscopes to include three, four, or more milk cartons to make the periscope longer or to give it creative shapes, as suggested in "For Problem Solvers." Additional mirrors might be needed as students expand with creative ideas. Other tubes could be substituted for the milk cartons. This is a good time to be creative.

Figure 3.11–1. Milk Carton Periscope

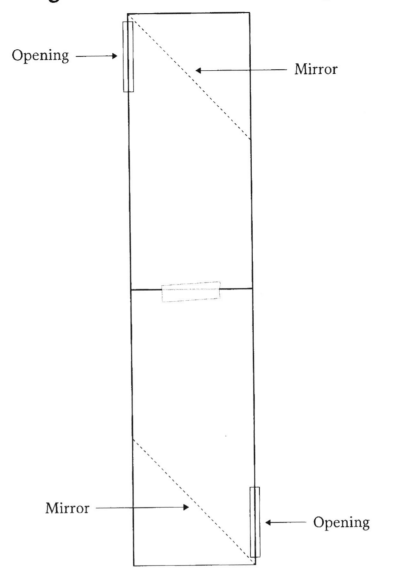

Opening

Mirror

Mirror

Opening

Integrating

Math

Science Process Skills

Observing, measuring, predicting, communicating, identifying and controlling variables, experimenting

Light

How Can You Pour Light?

(Enrichment activity)

Materials Needed

- Tall, slim olive jars with lids
- Flashlights
- Nails
- Masking tape or plastic tape
- Newspaper or light cardboard
- Hammer
- Water
- Sink or large bowl

Procedure

1. With the hammer and nail, make two holes in the lid of the jar. The holes should be near the edge but opposite each other. One hole should be quite small. Work the nail in the other hole to enlarge it a bit.

2. Fill the jar about two-thirds full of water and put the lid on, as seen in Figure 3.12–1.

Figure 3.12-1. Jar with Lid with Two Different-Sized Holes

3. Put tape over the holes in the lid until you are ready to pour.

4. Lay the jar and flashlight end-to-end, with the face of the flashlight at the bottom of the jar.

5. Roll the newspaper around the jar and flashlight to enclose them in a light-tight tube. Tape the tube together so it will stay. (See Figure 3.12–2.)

6. Slide the flashlight out of the tube, turn it on, and slide it back into the tube. Darken the room.

Figure 3.12–2. Flashlight and Jar Taped Together

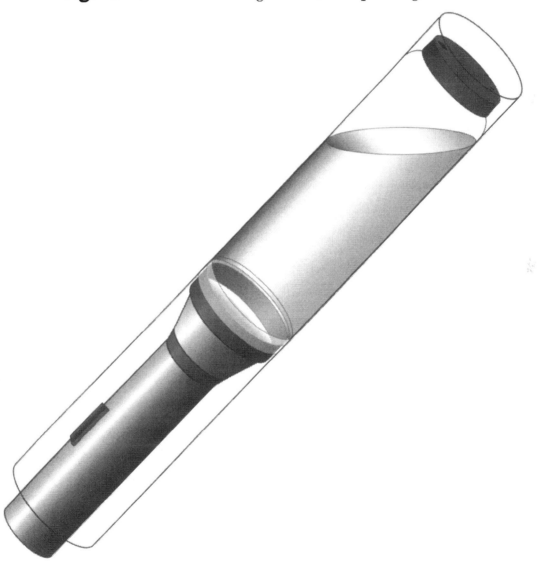

Figure 3.12-3. Water Pouring from Larger Hole

7. Hold the apparatus upright and remove the tape from the lid. With the large nail hole down, pour the water into the sink or large bowl, as shown in Figure 3.12-3.

8. What happened to the beam of light as the water poured into the sink or bowl?

9. Do you have any idea what caused this?

10. Share your ideas together.

For Problem Solvers

Use your creativity with this activity. Try different containers for the water and different light sources. Try putting some food coloring in the water. Does that provide the same effect as if you put a colored filter (colored acetate) over the flashlight?

Teacher Information

Although light travels in straight lines, it is reflected internally at the water's inner surface and follows the path of the stream of water. Because of the phenomenon of internal reflection, fiber optics can be used to direct light anywhere a wire can go, even into the veins and arteries of the human body.

Integrating

Language arts

Science Process Skills

Observing, inferring, communicating

What Color Is White?

(Take home and do with family and friends.)

Materials Needed

- White poster board
- Compass
- String 1 m (1 yd.) long
- Crayons
- Scissors

Procedure

1. With your compass, draw a circle 15 cm (6 in.) in diameter on the poster board.
2. Cut out the circle with the scissors.
3. Draw three equal pie-shaped sections on the poster board and color them red, green, and blue. (See Figure 3.13-1.)
4. Make two small holes near the center of the circle.
5. Thread the string through the holes (in one hole and out the other) and tie the ends of the string together, forming a loop that passes through the two holes of the disk.
6. Center the disk on the string loop and make the disk spin by alternately stretching and relaxing the string.
7. As the disk spins, watch the colored side.
8. What happens to the colors? Why do you think this happens?

For Problem Solvers

Obtain a copy of the "Science Investigation Journaling Notes" from your teacher, and try the same thing with more disks and different color combinations. Try using only two sections and coloring them with complementary colors, such as yellow and blue. Try several sections, alternating the same two colors back and forth and see whether you get the same result as with two large sections. Each time you try a new design or color combination, make a prediction of what you will see as the disk spins. Then try it, and test your prediction.

Hands-On Physical Science Activities

Figure 3.13-1. Divided and Colored Circle with Two Holes

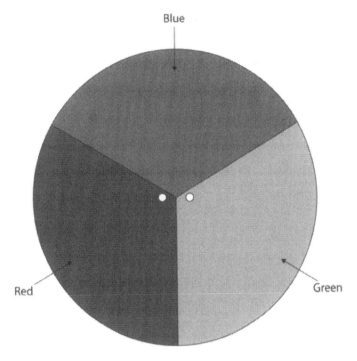

Teacher Information

As the primary colors spin, they should blend together to form a grayish white. If one of the colors seems to dominate, some of that color should be replaced with more of the other two colors. Blue might need a bit more than its share. Students will enjoy experimenting with various color combinations and testing their predictions of the resulting blends, as suggested in "For Problem Solvers."

The disk may spin better if its weight is increased by doubling the thickness of poster board or pasting the disk onto cardboard or by gluing a large button on the back of the disk. Another option is to mount the disk onto a sanding pad designed for an electric drill (use a drill with a variable-speed switch). The drill could then be used to spin the disk. *Caution:* If the drill is used at high speed, the disc could fly off and cause injury, so keep it toned down. You might want to use the drill as a teacher demonstration only.

Integrating

Art

Science Process Skills

Observing, inferring, classifying, measuring, predicting, communicating, comparing and contrasting, identifying and controlling variables, experimenting

How Can You Spin Different Colors?

Materials Needed

- White poster board
- Black fine-tipped markers
- Newspapers
- String 1 m (1 yd.) long
- Scissors
- Compasses
- Pencils
- Rulers

Procedure

1. Use a compass to draw a circle on the poster board 15 cm (6 in.) in diameter. Cut out the circle with the scissors, as seen in Figure 3.14–1.

2. Put a layer of several thicknesses of newspaper on the table and place your white disk on the newspaper.

3. With a fine-tipped marker, make one of the patterns shown in Figure 3.14–2. Let it dry.

4. Make two small holes on opposite sides of the center point, each about 1 cm (3/8 in.) from the center point, as in Figure 3.14–3.

5. Thread the string through the holes of the disk and tie the ends together.

6. Put one finger of each hand through the string, and wind up the disk on the string. Make the disk spin by successively pulling and relaxing the string.

7. As the disk spins, watch the painted side.

8. What do you see? Can you explain it?

For Problem Solvers

Create your own patterns for the disk. Yours might work out better than the ones shown here. Be creative.

Hands-On Physical Science Activities

Figure 3.14-1. Circle on Paper, Scissors

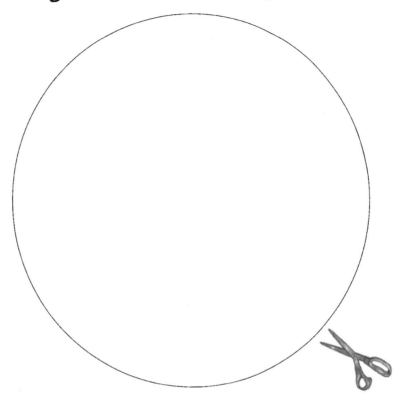

Figure 3.14-2. Disk Patterns

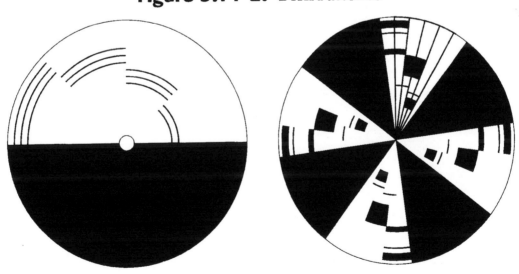

Figure 3.14-3. Center Point and Holes Marked on Patterns

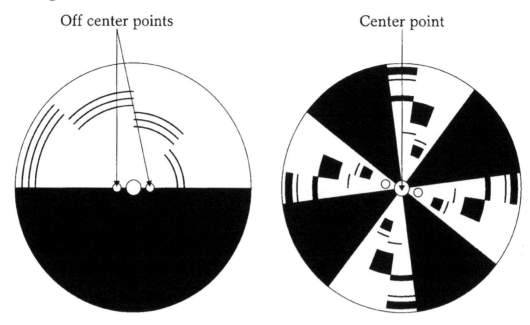

Teacher Information

When the retina receives repeated flashes of white light, they are interpreted by the brain as color. Such flashes of white light are produced by the spinning black-and-white disk. This phenomenon was discovered by Benham in the 19th century. If you have a phonograph turntable with adjustable speed, the effect can be studied by making a hole in the center of the disk and laying it on the turntable. Or attach the disk to a sanding pad for an electric drill, and turn the disk with the drill. Be sure the drill has a variable speed switch. *Caution:* If it spins too fast, the disk could fly off and cause injury.

Integrating

Art

Science Process Skills

Observing, inferring, classifying, measuring, predicting, communicating, comparing and contrasting, identifying and controlling variables, experimenting

What Do Color Filters Do to Colors?

(Whole-group activity)

Materials Needed

- A copy of the chart for Activity 3.15 for each student
- Red, blue, and green acetate, cut into strips about 5 cm (2 in.) by 15 cm (6 in.)
- Multicolored construction paper (one piece of each color available, each numbered with heavy black marker)

Procedure

1. Each person in the group should have one strip of red acetate, one strip of green, and one strip of blue.

2. Each person should have a copy of the chart on page 182.

3. Assign one person to be the paper holder.

4. Each person should hold the red acetate in front of his or her eyes.

5. The "paper holder" should hold up construction paper number 1.

6. Without removing the acetate from the eyes, each person looks at construction paper number 1 and writes the color he or she sees, without removing the acetate or trying to guess what color the paper really is.

7. The paper holder puts construction paper number 1 out of sight.

8. The paper holder holds up construction paper number 2, and each person again writes the color he or she sees.

9. Continue this procedure until all of the colors of construction paper have been used.

10. Repeat the entire process with participants using blue acetate.

11. Repeat the procedure again with green acetate.

12. Without looking at the construction paper, compare notes and predict what color each numbered paper really was.

13. Without acetate in front of eyes, the paper holder holds up construction paper number 1.

14. Discuss with the group what each person wrote as the color he or she saw when looking at paper number 1 through red acetate, blue acetate, and green acetate.

15. Repeat steps 12 and 13 with the other pieces of construction paper.

16. Discuss the effect of each color of acetate, and compare information from the group. Did everyone see the same color each time?

For Problem Solvers

Do some research on color blindness and see what you can learn about it. Do color-blind people see nothing but black and white? Do they see some colors but not others? If they lack only certain colors from their vision, what are the most common ones that are weak or missing for them? Use encyclopedias, the Internet, library books, and whatever resources you have available. Perhaps you know someone who claims to be color blind. Interview them; they are the ones who really know how it is for them. Is color blindness the same for everyone who claims to be color blind?

Compare your results with that of others who do this activity, and together share your information with the class.

Teacher Information

We see objects because of the light they reflect. If no red is present in the color that is reflected from the paper, and the acetate filters out everything but red, the person viewing the paper will probably see an unpleasant muddy color.

The strip of acetate can be turned into a pair of groovy goggles and add interest to the activity. Staple the acetate to a two-hole section from a plastic six-pack can carrier, as shown in Figure 3.15–1. Rubber bands can then be attached to the ends of the goggles. The rubber bands can go over the ears or on around the back of the head to hold the goggles in place. Students might do other things to add interest and design to the goggles.

Other items around the room can be viewed with the goggles. If there is something of a fluorescent color on the wall of the classroom or on someone's T-shirt, for instance, great aesthetic experiences will be had by all!

This activity often sparks a great interest in color blindness and the reasons we see colors. We describe "color" as a characteristic of an object or a scene, but is it really? Our eyes respond to stimuli from different wavelengths of light, and send messages to the brain. The interpretation of those messages results in mental perceptions of color for us. In our world of modern technology we should appreciate the very complex "computer" equipment in our heads that cause this to happen and help us to appreciate the world around us.

Hands-On Physical Science Activities

This would be a great time to study color blindness. Students who do the "For Problem Solvers" will be prepared to lead a discussion that could lead to other activities and other discussions, as far as you want it to go, according to the interest of your class and the time allowed for it.

Integrating

Art

Science Process Skills

Observing, inferring, classifying, measuring, predicting, communicating, comparing and contrasting

Figure 3.15-1. Six-Pack Goggles

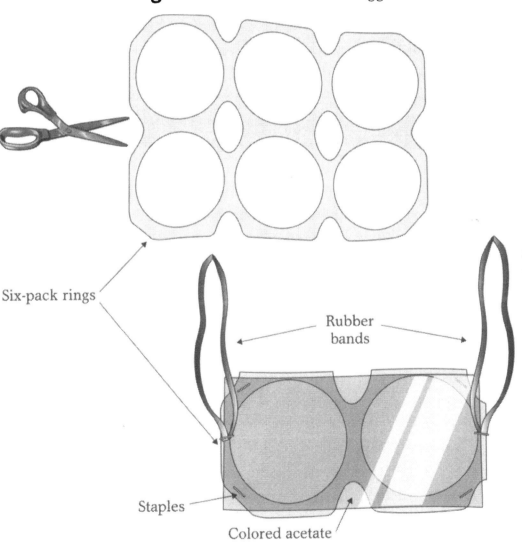

Chart for Activity 3.15

Construction Paper Number

Color of Acetate	1	2	3	4	5	6
Red						
Blue						
Green						

ACTIVITY 3.16

How Do Detectives Use Color to Solve Crimes?

(Take home and do with family and friends.)

Materials Needed
- Water-based markers
- Coffee filters (or other filter paper)
- Half-liter (pint) jars
- Scissors
- Water

Procedure
1. Cut the filter paper into strips, about 2.5 cm (1 in.) wide.
2. Using one of the markers, place a dot about 2.5 cm (1 in.) from one end of one strip of filter paper.
3. Put a small amount of water in the jar.
4. Place the strip of filter paper inside the jar, folding the top of the strip over the lip of the jar to hold it in place. (Anchor the filter paper in place with a paper clip if necessary.) Be sure the end of the filter paper is in the water and that the colored dot is above the water. (See Figure 3.16–1.)
5. Write your prediction of what will happen to the color as the water soaks up through the filter paper.
6. Check the filter paper every few minutes for about thirty minutes.
7. Was your prediction accurate?
8. Discuss your observations. Try to explain what happened.

Figure 3.16-1. Strip of Filter Paper in Cup

For Problem Solvers

Obtain a copy of the "Science Investigation Journaling Notes" from your teacher, and extend this activity to more than just water-based markers. Using water as the solvent, you can use a powdered drink mix (use only a small amount of water to make the dot, so it is highly concentrated), and food coloring. For permanent ink you can try white vinegar or rubbing alcohol as the solvent. Try various brands of markers—any you can find. Find out if all manufacturers use the same color combinations in making specific colors. Keep a record of your findings and share your information with others who are doing this activity, or who are interested in it.

If you had a friend who was allergic to a certain color of food dye, could you use this technique to find out whether that color is in a particular package of powdered drink mix?

How do you think police investigators might use this technique to help solve the mystery of a crime?

Teacher Information

This process is called "color chromatography." Many inks used in pens and markers have surprising combinations of coloring agents in them, even black ink. This is also true of coloring agents used in powdered drink mixes and in food colorings. Colors that are soluble in water will dissolve into the water as the water makes its way up the filter paper. How high the color will go up the paper will depend on how soluble that particular ingredient is and on how well it binds or sticks to the filter paper.

Color chromatography is actually used by investigators in solving mysteries. For example, could the pen found on Joe Scribbler's body have been used in writing the suicide note? If the color chromatography technique reveals that both the ink on the note and the ink in the pen contain the same combination of coloring agents, it *might have been* the pen used for the note. If the color ingredients are different, it *was not* the pen used.

Integrating

Art, social studies

Science Process Skills

Observing, inferring, classifying, measuring, predicting, communicating, comparing and contrasting, using space-time relationships, formulating hypotheses, identifying and controlling variables, experimenting

What Must You Remember When Spearing a Fish?

Materials Needed
- Plastic dishpans
- Coins, paper clips, or other metal objects
- Meter sticks (or yardsticks)
- Tape
- Metal rods (a straightened coat hanger will do)

Procedure
1. Fill the dishpan about two-thirds full with water.
2. Place a coin at the bottom of the pan. (This is your "fish"!)
3. Make a V-shaped trough by taping the two meter sticks together.
4. Hold one end of the trough at the edge of the pan, *above the water,* and try to spear the "fish" by aiming with the trough and sliding the metal rod down the trough.
5. Did you hit the fish? How close did you come?
6. What happened? Did the fish swim away? Do you need glasses?
7. Practice and see whether you can improve your accuracy.
8. Discuss what you had to do to hit the fish. Talk about your ideas.

Teacher Information
If students keep the end of the trough out of the water, they will most likely miss the "fish" considerably on their first try. With practice they will learn that the object is not where it appears to be, and they will compensate and increase their accuracy. This experience will provide an effective introduction to the concept of refraction of light and the optical illusions that refraction causes.

Talk about eagles and bears and their ability to strike at the right place when catching a fish. How do you think they learned to do that? Perhaps eagles and bears also learn by trial and error.

Integrating
Math, language arts

Science Process Skills
Observing, inferring

What Happened to the Pencil?

(Take home and do with family and friends.)

Materials Needed

- Clear tumbler or bowl
- Water
- Pencil (or spoon)

Procedure

1. Fill the tumbler or bowl about two-thirds full of water.
2. Put the pencil into the water.
3. Look at the pencil from the top and from the side.
4. What appears to happen to the pencil at the water level? What ideas do you have about this effect?

Teacher Information

Light travels at different speeds through different substances, creating a bending effect on any light rays that enter a substance at an angle. This is called refraction. Light travels faster through air than it does through water. The bending of the light rays as they pass from air to water or from water to air results in an optical illusion as the object in the water appears to be broken at the surface of the water.

Integrating

Math, language arts

Science Process Skills

Observing, inferring, predicting, researching

Can You Find the Coin?

(Take home and do with family and friends.)

Materials Needed

- Opaque bowl
- Water
- Coin (or button)

Procedure

1. Place the coin in the bowl.
2. Stand in such a position that the coin is just hidden from your view by the edge of the bowl.
3. Without shifting your position, have a partner slowly fill the bowl with water, being careful not to disturb the coin at the bottom of the bowl.
4. What happened to the coin as your partner poured water into the bowl?
5. What do you think could have caused this? Share your observations and ideas with your family and with others who are doing the same activity.

Teacher Information

Light travels in what appears to be a straight line in air, but when it passes from water to air, it is bent by refraction, because it travels more slowly through water than through air. As water is poured into the bowl, the light will bend and more of the bottom of the bowl will be exposed. The coin will appear.

Bowl and Coin Showing How Water Bends Light

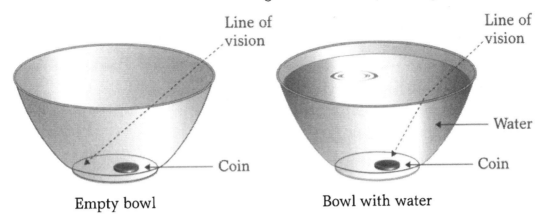

Empty bowl

Bowl with water

Integrating

Language arts

Science Process Skills

Observing, inferring, predicting, communicating, comparing and contrasting, identifying and controlling variables

How Can a Postage Stamp Hide Under Clear Glass?

(Take home and do with family and friends.)

Materials Needed

- Empty short jar (such as peanut butter jar) with lid
- Water
- Postage stamp or sticker

Procedure

1. Put the stamp on a table.
2. Fill the jar with water and put the lid on.
3. Place the jar on the stamp.
4. Look at the stamp.
5. Explain your observations.

For Problem Solvers

Replace the jar with a plastic cup. Try this same activity with many different containers. They all need to be clear, of course, but try various shapes and sizes, using both glass and plastic. Do they all work the same? What are the differences? Can you tell why?

Teacher Information

As light passes from water to air, the light bends (refracts) because it travels through these materials at different speeds. In this activity, the refraction makes the stamp appear higher than it really is. When it is looked at from an angle, reflected light from the stamp doesn't reach the eyes, so the stamp seems to have disappeared. The lid on the jar prevents the observer from looking straight down on the stamp.

Integrating

Language arts

Science Process Skills

Observing, inferring, predicting, communicating

What Makes Light Bend?

Materials Needed

- Aquarium three-fourths full of water
- Flashlights or projector (or laser pointers)
- Milk
- Two chalky chalkboard erasers per small group
- One sheet of paper per student
- Tape
- Copies of "Science Investigation Journaling Notes" for all students

Procedure

1. Do the following as a Science Investigation. Obtain a blank copy of the "Science Investigation Journaling Notes" from your teacher. Write your name, the date, and your question at the top. Plan your investigation through item 5 of the form (Procedure) and have it approved by your teacher. Complete the Journaling Notes as you perform your investigation. When you are finished, share your project with your group and submit your Journaling Notes to your teacher if requested.

2. Pour milk into the aquarium, a little at a time, until the water has a slightly cloudy appearance. Easy does it. You might need only a spoonful.

3. Wrap and tape the paper around the flashlight like a tube, to focus the light into a narrow beam.

4. Turn the room lights off.

5. Aim the light at the water on an angle. Have someone clap the chalkboard erasers together over the aquarium to make the beam of light easier to see in the air, as in Figure 3.21–1.

6. Observe carefully the angle of the light beam as it extends from the flashlight to the water and as it continues through the water.

7. What happens to the light beam as it enters the water? What do you think causes the change?

8. Complete your "Science Investigation Journaling Notes." Are you ready to explain what makes light bend? Discuss it with your group.

Figure 3.21-1. Setup for Activity 3.21

Teacher Information

This experiment makes the phenomenon of refraction easily visible. With the chalk dust, the light can be seen in the air, and the milk makes the change in direction easily observable in the water.

Caution: The laser pointer is more effective with this activity than is the flashlight or projector because of the well-defined beam of light that it produces. If the laser pointer is used, however, extreme caution must be taken to see that it is not pointed at the eyes.

Integrating

Math, language arts

Science Process Skills

Observing, inferring, comparing and contrasting

How Can You Make a Glass Disappear?

Materials Needed

- Two large glass jars per small group
- Two small glass jars or drinking glasses per small group
- Water
- Cooking oil
- Copies of the "Science Investigation Journaling Notes" for all students

Procedure

1. Place the two small jars inside the large jars.
2. Fill one pair of jars with water.
3. Can you see the small jar?
4. Fill the other pair of jars with cooking oil.
5. Can you see the small jar?
6. Explain your observations.

For Problem Solvers

Obtain a copy of the "Science Investigation Journaling Notes" from your teacher, and think of some other ways to do this activity. Use containers with different shapes. Try plastic containers instead of glass containers. Does it work out differently in water if you put food coloring in the water? Think of other ways to test refraction of light.

Teacher Information

As light passes from one transparent material to another (such as air, water, and glass), the light is bent at the boundary between the two materials. This happens because of the differing speeds at which the materials transmit light. Light moves at about the same speed through petroleum products (including cooking oil) as it does through glass. Therefore, as light passes between glass and oil it doesn't bend at the boundaries, leaving the boundaries invisible. The inner jar in water will likely be visible, whereas the inner jar in cooking oil likely will not.

Integrating

Math, language arts

Science Process Skills

Observing, inferring, communicating, comparing and contrasting, identifying and controlling variables

How Does a Camera See the World?

(Teacher demonstration)

Materials Needed

- Pinhole camera (see Figure 3.23-1)
- Candle
- Match

Procedure

1. Darken the room.
2. Light the candle.
3. Point the pinhole in the box toward the candle.
4. Have students look at the image on the tissue paper at the back of the box. Ask what they observe about the image of the candle flame.
5. Ask, "What can you say about this?"

Teacher Information

As the light from the candle passes through the pinhole, the image is inverted because light travels in a straight line. (See Figures 3.23-2 and 3.23-3.)

Figure 3.23-1. Shoe Box with Pinhole and Tissue Paper

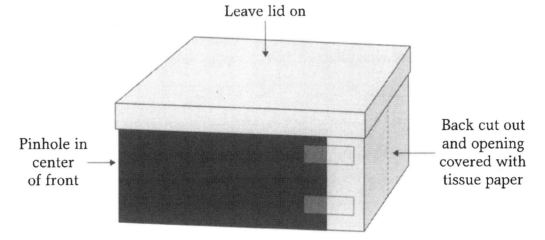

Leave lid on

Pinhole in center of front

Back cut out and opening covered with tissue paper

Figure 3.23-2. Candlelight Going Through Pinhole Camera

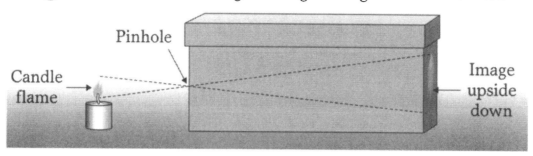

Figure 3.23-3. End View of Box

The human eye also receives images upside down on the retina, but the brain somehow turns them right side up again as we "see" them. *Caution:* Close supervision of candle flame is needed. This should be a teacher-demonstration activity.

Integrating
Language arts

Science Process Skills
Observing, inferring

How Does a Lens Affect the Way Light Travels?

(Teacher-supervised activity)

Materials Needed
- Candles
- Matches
- White cardboard
- Hand lenses
- Pans

Procedure
1. Prop a piece of the cardboard on a table.
2. Stand a candle in a pan or other nonflammable container about 60 to 90 cm (2 to 3 ft.) away from the cardboard.
3. Light the candle.
4. Hold the hand lens near the cardboard. Move it slowly toward the flame until a clear image of the flame appears on the cardboard. (See Figure 3.24–1.)
5. Do you see anything strange about the flame? What effect do you think the hand lens has on what you see?
6. Share your observations and ideas with your group.

Figure 3.24–1. Candle, Lens, and Cardboard

For Problem Solvers

Try to find a variety of lenses, different types and sizes. Try the candle-flame activity with each lens. If you notice differences in what happens, describe those differences. If you have many lenses, put them in groups according to the way they worked for you.

Teacher Information

The bending of light through refraction results in an inverse image of the flame as it is projected onto the cardboard. The same thing happens with the eye. Images are projected onto the back of the eye upside down, but they are reversed to their true perspective as they are interpreted by the mind.

Integrating

Math, language arts

Science Process Skills

Observing, inferring

How Can You Make a Lens from a Drop of Water?

(Take home and do with family and friends.)

Materials Needed

- Small sheet of clear plastic or glass (or even plastic wrap)
- Eyedropper
- Water
- Book

Procedure

1. Place a drop of water on the sheet of plastic.
2. Lay the plastic over the page of a book and look at a letter or punctuation mark through the drop of water.
3. What does the drop of water do to the images on the paper?
4. Examine other things through the drop of water, such as a piece of cloth or the back of your hand, by laying the plastic on them.
5. Talk to each other about what happened and why.

For Problem Solvers

Do this project as a Science Investigation. Obtain a blank copy of the "Science Investigation Journaling Notes" from your teacher. Write your name, the date, and your question at the top. Plan your investigation through item 5 (Procedure) and have it approved by your teacher. Complete the Journaling Notes as you perform your investigation. Share your project with your group, and submit your Journaling Notes to your teacher if requested.

Get a piece of wire that you can bend easily and make a small loop in one end, just about the width of a drop of water. Place a drop of water in the loop. Hold the loop near a page of print. What happens? How is it different from what you saw with the water on plastic in the above activity? Try making a wire loop just a little bit larger; then make one a little bit smaller. Does it seem to make any difference? Shake some of the water out of the wire loop, leaving just enough to remain stretched across the loop. Look at the page of print again. What do you think makes the difference?

Teacher Information

Any transparent substance with a convex surface will cause light rays to bend and converge. Many vision-aiding devices are based on this principle, including eyeglasses, hand lenses, binoculars, microscopes, and telescopes. The drop of water isn't the best lens, because it must be handled carefully and the degree of surface curve is difficult to control, but it is a lens.

If you prefer, you can bypass the sheet of plastic and place the drop of water directly on a page of print. The plastic, however, provides transportability for multiple uses.

Integrating

Language arts

Science Process Skills

Observing, inferring, comparing and contrasting

How Are Convex and Concave Lenses Different?

Materials Needed

- Convex lenses
- Concave lenses
- Small sheets of clear glass
- Flashlights or projector (or laser pointer)
- Sheets of paper
- Tape
- White surface

Procedure

1. Roll the paper into a tube around the end of the flashlight and tape it in place, as shown in Figure 3.26–1.
2. Shine the light on a white surface.
3. Hold each lens and the sheet of glass, one at a time, in the path of the light. What happened each time? Share your ideas with each other.

For Problem Solvers

Examine the surface of a book, a picture, your fingernail, and other objects using the two lenses. Explain the differences in the images you see. Examine the shapes of the lenses carefully. How are they alike? How are they different?

Figure 3.26–1. Flashlight with Paper Tube Around It

Figure 3.26-2. Convex and Concave Lenses

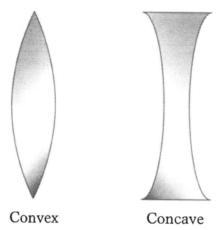

Convex   Concave

One of these lenses is called a concave lens and the other is called a convex lens, as shown in Figure 3.26–2. Do some research about lenses and see whether you can find out why they have a different effect on images that are seen through them.

Teacher Information

When light passes between media of differing densities (such as air and glass or air and water), the light can be refracted, or bent. Convex lenses are thicker in the middle than on the edges and cause light rays to converge, or come together. Concave lenses are thicker on the edges than in the middle and cause light rays to diverge. A convex lens in the path of the light will concentrate the light on the white surface, causing it to appear brighter, while a concave lens will spread the beam of light over a larger surface. Convex lenses are used as magnifiers, while concave lenses are used to make things appear smaller.

Caution: The laser pointer is more effective with this activity than is the flashlight or projector because of the well-defined beam of light that it produces. If the laser pointer is used, however, extreme caution must be taken to see that it is not pointed at the eyes.

Integrating

Language arts

Science Process Skills

Observing, inferring, comparing and contrasting

How Can You Measure the Magnifying Power of a Lens?

Materials Needed
- Hand lenses
- Lined paper

Procedure
1. Lay the hand lens on the lined paper and count the number of lines from one edge of the lens to the other.
2. Pick up the lens and hold it in such a position that the lines on the paper come into focus.
3. How many lines do you see in the lens?
4. Compare the number of lines in step 1 with the number of lines in step 3.
5. From your information, what would you say is the magnifying power of your lens: Two power? Three power? Five power?

Teacher Information
If the lens itself spans six lines but only three lines can be seen through the lens when held in focus position, the lens is about two power. If only two of the six can be seen through the lens in focus position, the lens is about three power. It makes things look about three times as large as they are. This is not an accurate measurement of lens magnification, but it will provide a close estimate. If you are using a ten-power lens, you will need to use lines that are quite close together in order to get a workable count. The smaller and more powerful the lens, the closer the counted lines will need to be.

Integrating
Math

Science Process Skills
Observing, measuring, predicting, communicating, comparing and contrasting

What Does a Prism Do to Light?

Materials Needed

- Copies of "Science Investigation Journaling Notes" for this activity for each student
- Prism
- Projector or flashlight
- Screen, white paper, or white wall

Procedure

1. As you complete this activity you will keep record of what you do, just as scientists do. Use the form "Science Investigation Journaling Notes" and write the information that is called for, including your name and the date.

2. For this activity you will learn what a prism does to light. For item 1, the question is provided for you on the form.

3. Item 2 asks for what you already know about the question. If you have some ideas about prisms and light, write your ideas.

4. For item 3, write a statement of what a prism does to light, according to what you already know, and that will be your hypothesis.

5. Now continue with the following instructions. Complete your Journaling Notes as you go. Steps 6 through 10 below will help you with the information you need to write on the form for items 4 and 5.

6. Shine the light on a screen or other white surface.

7. What do you think will happen if you place the prism in the path of the beam of light?

8. Place the prism in the path of the beam of light. Were your predictions accurate?

9. What is white light?

10. Which color seems to bend the most as light passes through the prism? Which bends least?

11. Complete your "Science Investigation Journaling Notes." Are you ready to explain what a prism does to light? Discuss it with your group.

Hands-On Physical Science Activities

For Problem Solvers

Examine the colors that you get as the light shines from the projector through the prism and onto the white background very carefully. This is called a color spectrum. How many colors are there? Make a sketch of the color bands on paper and give each color a name. Then do the same thing again, using a flashlight instead of a projector. Do you get the same colors? Are they in the same order? Are they as bright? Try it again, using the sun as your light source. Try it with a light bulb. Try a laser pointer.

Can you find still other light sources to project light through your prism? Are the colors and the sequence always the same? Did you try a colored light bulb? Try shining the light through colored plastic film before it reaches the prism. Does that change anything?

Share your findings with others who are doing this investigation.

Teacher Information

White light is a combination of many colors. Each color has its own wave length and is bent to a different degree as light passes through a prism, forming a continuous spectrum. Violet has the shortest waves and is bent the most. Red has the longest waves and is bent the least. Five other "pure" colors exist in the spectrum between violet and red. In order, these are indigo, blue, green, yellow, and orange. Sometimes indigo and blue are treated as one color.

Water droplets in the air can act as tiny prisms when conditions are right, thus creating a rainbow. Many jewelry stores sell leaded glass crystals cut in different shapes; these crystals act as prisms and produce beautiful rainbows.

Integrating

Art

Science Process Skills

Observing, inferring, classifying, predicting, communicating, comparing and contrasting, formulating hypotheses, identifying and controlling variables, experimenting

Science Investigation

Journaling Notes for Activity 3.28

1. Question: *What does a prism do to light?*

2. What we already know:

_____

_____

3. Hypothesis:

_____

_____

4. Materials needed:

_____

_____

5. Procedure:

_____

_____

6. Observations/New information:

_____

_____

7. Conclusion:

_____

_____

How Can You Make a Prism with Water?

(Take home and do with family and friends.)

Materials Needed

- Sunny day
- Mirrors
- Trays or pans
- Water
- White surface

Procedure

1. Place the tray on a table or on the floor in direct sunlight. Put about 2 to 3 cm (1 in.) of water in the tray.

2. Place a mirror in the tray and focus the reflected sunlight on the white surface. At least part of the mirror should be submerged in the water. Try different depths of water and angles of the mirror. (See Figure 3.29–1.)

3. What do you see in the reflection on the white surface?

Figure 3.29-1. Setup for Activity 3.29-1

For Problem Solvers

If you have an empty aquarium, put it in a place in a room where sunlight is coming through a window and will shine directly on the aquarium. Put several inches of water in it and place a large mirror at the bottom, standing at an angle and leaning against the side. With the sun shining into the mirror, look around the room for rainbows. Look directly into the aquarium and walk around it slowly, again looking for rainbows.

Teacher Information

As light passes through a transparent substance (glass, plastic, water) at an angle, the light rays are bent. White light contains many other colors, each of which bends to a different degree. Thus, the reflected light on the wall shows a separation of those colors. The colors separate as light passes through a glass prism, and the same effect is produced in this activity with a water prism.

This same phenomenon occurs in nature as water droplets in the air separate the colors in sunlight. We see it as a rainbow. Rainbows can sometimes be seen in fine sprays of water, such as that produced by some lawn sprinklers.

Integrating

Art

Science Process Skills

Observing, inferring, classifying, predicting, communicating, formulating hypotheses, identifying and controlling variables, experimenting

Can You Solve This Light Word Search?

Try to find the following Light terms in the grid below. They could appear in horizontal (left to right), vertical (up or down), or diagonal (upward or downward) position.

shadow	light	dark
sun	concave	energy
magnify	reflect	refraction
convex	absorb	transparent
translucent	opaque	color

```
M  N  B  C  O  N  V  E  X  M  N  F
R  F  V  C  O  L  L  I  G  H  T  C
E  D  R  E  F  L  E  C  T  N  Y  X
F  G  U  D  O  P  O  C  E  Z  T  D
R  T  J  A  B  S  O  R  B  M  R  E
A  R  K  R  O  P  A  Q  U  E  C  W
C  Y  E  K  W  P  S  H  A  D  O  W
T  R  A  N  S  L  U  S  C  E  N  T
I  S  U  N  E  O  P  L  K  B  C  U
O  Y  A  R  E  R  M  X  C  E  A  Y
N  R  Y  U  K  L  G  K  J  R  V  A
T  M  A  G  N  I  F  Y  U  Y  E  R
```

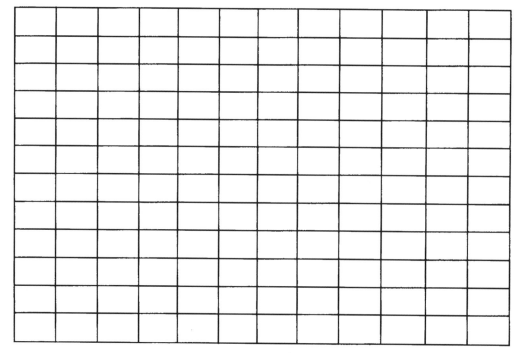

Can You Create a New Light Word Search of Your Own?

Write your Light words in the grid below. Arrange them in the grid so they appear in horizontal (left to right), vertical (up or down), or diagonal (upward or downward) position. Then fill in the blank boxes with other letters. Trade your Word Search with someone else who has created one of his or her own, and see whether you can solve the new puzzle.

Answer Key for Light Word Search

M	N	B	**C**	**O**	**N**	**V**	**E**	**X**	M	N	F
R	F	V	C	O	L	**L**	**I**	**G**	**H**	**T**	C
E	D	**R**	**E**	**F**	**L**	**E**	**C**	**T**	N	Y	X
F	G	U	**D**	O	P	**O**	C	**E**	Z	T	D
R	T	J	**A**	**B**	**S**	**O**	**R**	**B**	M	**R**	E
A	R	K	**R**	**O**	**P**	**A**	**Q**	**U**	**E**	**C**	W
C	Y	**E**	**K**	W	**P**	**S**	**H**	**A**	**D**	**O**	**W**
T	**R**	**A**	**N**	**S**	**L**	**U**	**S**	**C**	**E**	**N**	**T**
I	**S**	**U**	**N**	**E**	O	P	L	K	B	**C**	U
O	Y	**A**	R	E	**R**	M	X	C	E	**A**	Y
N	R	Y	U	K	L	**G**	K	J	R	**V**	A
T	**M**	**A**	**G**	**N**	**I**	**F**	**Y**	U	Y	**E**	R

Do You Recall?

Section Three: Light

1. Shadows are often darker toward the middle than around the edges. Why is that?

2. When you are outdoors, why does your shadow change as the day moves on?

3. Does light always travel in a straight line?

4. Write two examples of source light.

5. How does the angle of a beam of light approaching an object compare with the angle of the light reflecting from the object?

Do You Recall? *(Cont'd.)*

6. What color do you see when you look at an object that reflects all colors?

7. What color do you see when you look at an object that reflects no colors at all?

8. Consider a pencil standing in a cup of water and leaning against the side of the cup. Why does the pencil appear to bend?

9. When the light of a candle passes through a pinhole, or through a lens, and onto a screen, the image of the flame is upside down. Why does that happen?

10. Why does a drop of water magnify things?

Light

Do You Recall? *(Cont'd.)*

11. Do concave lenses make things appear larger or smaller?

12. Do convex lenses make things appear larger or smaller?

13. What does a prism do to light?

Answer Key for Do You Recall?

Section Three: Light

Answer   **Related Activities**

1. Depending on the relative size and position of the light source and the object blocking the light, the light is not fully blocked from the area at the edge of the shadow.   3.1

2. Shadows change continuously as the sun moves across the sky, constantly adjusting the angle of the light source.   3.2

3. Yes, it can be reflected but it always moves in a straight line.   3.3

4. Sun, fire, electric lights, fireflies (give off their own light)   3.6

5. They are the same.   3.7–3.10

6. White   3.13, 3.14

7. Black   3.13, 3.14

8. Light travels through water and air at different speeds, and is bent when it passes from one to the other.   3.17–3.22

9. Because light travels in a straight line   3.23, 3.24

Answer	**Related Activities**
10. This is due to refraction, or the bending of light as it moves from one medium to another. Any clear, domed surface tends to magnify.	3.17–3.22, 3.25
11. Smaller	3.26
12. Larger	3.26
13. Separates the colors	3.28, 3.29

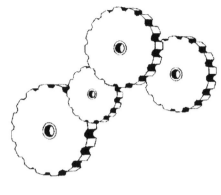

Sound

To the Teacher

Sound is a very important part of our lives. It is one of the first stimuli to which newborn infants respond, and its presence or absence shapes and affects us throughout our entire lives. This section introduces sound, its causes, and its uses. A study of sound lends itself very well to concrete activities, with many possibilities for discovery/inquiry. No attempt is made to introduce the physiology of the ear, although you may choose to teach it in relation to this area.

A study of sound can be greatly expanded in the Language Arts to develop and enrich listening skills. Music can be integrated through discussion of musical terms found within this section. Activities on inventing and playing musical instruments could lead to additional study of ancient methods and modern electronic methods of producing music.

Throughout the study, children should be encouraged to bring and demonstrate their own musical instruments. If there is a high school or university nearby, the music director may be willing to provide musicians and instruments. Most communities have choral groups that might be willing to perform. A note to parents asking for the names of family members or neighbors who play unusual musical instruments could produce interesting and entertaining results.

As the importance of sound is discussed, the value of being able to hear and speak clearly should be emphasized. Children should know that people of all ages suffer from hearing loss, and that almost everyone develops some degree of impairment as he or she grows older.

Sometime near the beginning of the study, the class should discuss, and perhaps list, ways sound can help us; for example, communication, warning, entertainment, aesthetics, and protection.

Sound can also be harmful. Loud noise can injure the ears. Sound can be pleasing and soothing to an individual, but it can also be disturbing and irritating.

The loudness of sound is measured in *decibels.* For public protection, many communities have laws restricting the decibel levels that can be produced by any means. With electronic sound equipment being so common, children should understand reasons for attempts to control noise levels.

In discussing *pitch,* or the frequency of vibrations, children should be aware that the human ear cannot detect the frequency of very high (fast) and very low (slow) vibrations. Dog whistles are pitched too high to be heard by people, but dogs and some other animals can hear them.

Resource people can be involved frequently in this study. These might include individuals of all ages who have hearing handicaps, a nurse, a doctor, an audiologist, an acoustics specialist, the manager of a music store, a musician, someone who makes or plays unusual musical instruments, or any others you may find helpful.

This is an area rich in "take home and do with family and friends" activities. Taking home concrete objects to show and talk about will help children develop increased language ability and be a source of strong motivation in science.

The following activities are designed as discovery activities that students can usually perform quite independently. You are encouraged to provide students (usually in small groups) with the materials listed and a copy of the activity from the beginning through the "Procedure." The section titled "Teacher Information" is not intended for student use, but rather to assist the teacher with discussion following the hands-on activity, as students share their observations. Discussion of conceptual information prior to completing the hands-on activity can interfere with the discovery process.

Regarding the Early Grades

With verbal instructions and slight modifications, many of these activities can be used with kindergarten, first-grade, and second-grade students. In some activities, steps that involve procedures that go beyond the level of the child can simply be omitted and yet offer the child an experience that plants the seed for a concept that will germinate and grow later on.

Teachers of the early grades will probably choose to bypass many of the "For Problem Solvers" sections. That's okay. These sections are provided for those who are especially motivated and want to go beyond the investigation provided by the activity outlined. Use the outlined activities and enjoy worthwhile learning experiences together with your young students. Also consider, however, that many of the "For Problem Solvers" sections can be used appropriately with young children as group activities or as demonstrations, still giving students the advantage of an exposure to the experience and laying groundwork for connections that will be made later on.

Correlation with National Standards

The following elements of the National Standards are reflected in the activities of this section.

K-4   Content Standard A: Science as Inquiry
As a result of activities in grades K-4, all students should develop:
  1. Abilities necessary to do scientific inquiry
  2. Understanding about scientific inquiry

K-4   Content Standard B: Physical Science
As a result of activities in grades K-4, all students should develop understanding of
  1. Properties of objects and materials
  2. Position and motion of objects
  3. Light, heat, electricity, and magnetism

5-8   Content Standard A: Science as Inquiry
As a result of activities in grades 5-8, all students should develop:
  1. Abilities necessary to do scientific inquiry
  2. Understanding about scientific inquiry

5-8   Content Standard B: Physical Science
As a result of activities in grades 5-8, all students should develop understanding of
  1. Properties and changes of properties in matter
  2. Motions and forces
  3. Transfer of energy

Sound

How Are Sounds Produced?

Materials Needed

- Meter sticks (or yardsticks)
- Foot rulers
- Tongue depressors
- Partners

Procedure

1. Place the meter stick on a table, with part of it extending over the edge of the table.

2. Hold the stick firmly on the table while your partner flips the other end of the stick (or you can hold the stick with one hand and flip it with the other).

3. Repeat this process with different lengths of the stick extending over the edge of the table.

4. What happens to the sound as the vibrating part of the stick is longer and shorter?

5. Does the stick seem to vibrate faster or slower when the vibrating part of the stick is longer? What happens to the pitch?

6. Does the stick seem to vibrate faster or slower when the vibrating part of the stick is shorter? What happens to the pitch?

7. Try these same actions with the tongue depressor and the foot ruler. Do they produce the same sounds? Do the sounds change in similar ways as the vibrating part becomes longer or shorter?

For Problem Solvers

Continue this investigation by using different types of sticks and other materials. Predict the result before you try each new object. Discuss your observations with your group.

Teacher Information

Emphasize these basic science concepts: (1) If a sound is heard, something is vibrating, and (2) when the vibrating portion of the object is longer, the object vibrates more slowly and the pitch goes down; and when the vibrating portion is shorter, the vibration is faster and the pitch goes up. This is an excellent activity for helping students to associate sounds with vibrating objects, because the vibrations can be detected visually and easily associated with the changes in sound as the variables are manipulated. Students will observe that when the vibrating portion of the object is longer the pitch goes down, and when the vibrating portion is shorter the pitch goes up. Be sure to reinforce the connections between their observations and the science concepts involved.

Integrating

Language arts

Science Process Skills

Observing, inferring, predicting, controlling variables

What Sounds Can You Identify?

(Whole-group activity)

Materials Needed

- Cassette player
- Tape of common sounds
- Pencils and paper

Procedure

1. Listen to the first sound your teacher will play. Write what you think you heard. Do this for each sound on the tape.

2. Name and describe as many of the sounds as you can.

3. Use descriptive words, such as high, low, squeaky, or gruff, as you discuss the differences you hear in the sounds. Think of other words to describe sound.

4. Discuss the effects that sounds have in your life.

For Problem Solvers

If you have a tape recorder at home, record some sounds that you hear at home. Try to find or create some unusual sounds. Then take your tape to school and see whether your classmates can identify the sources of the sounds you recorded.

Teacher Information

Many teachers enjoy preparing tapes of different sounds. You may want to prepare one on everyday sounds, such as a phone ringing, water running, dog barking, bird chirping, car starting, animal noises (sound toys available for young children are good sources), automobile horns, musical instruments, or jet airplane taking off. And yes, the sound of a flushing toilet is a must for such a recording; it is sure to get a response. Many schools have sound records for use with kindergarten and primary grades. Check your media center or library.

Take a sound field trip. Take several tape recorders on a field trip and collect as many sounds as you can.

Encourage children to do the "For Problem Solvers" activity and see whether they can stump their classmates with sounds they hear at home.

Play soft classical music and loud rock music. Have children make a painting or color a picture showing how each makes them feel.

Classify sounds into categories; for example, warning sounds: siren, bell, honking horn, growling dog, screeching brakes. (Mothers' voices as they continue to call children often change in interesting ways.)

Have a class discussion of how sounds help us and occasionally harm us (too loud may damage ears or make us nervous).

Integrating

Language arts, art, music

Science Process Skills

Observing, inferring, classifying, communicating, using space-time relationships

How Well Can You Match Sounds?

(Take home and do with family and friends.)

Materials Needed

- Set of film canisters, each containing different small objects
- Masking tape

Procedure

1. Shake the canisters and listen to the noise they make.
2. Can you hear the different sounds they make?
3. Do two or more of the canisters make the same sound?
4. If you find canisters that sound alike, put them next to each other.
5. Have a friend listen to the canisters and see whether he or she agrees.
6. You may want to make more rattle cans with different sounds to see how well your friends can detect the differences in the sounds.

For Problem Solvers

Make your own set of rattle cans. Put small things in them. You can use any containers that have lids and that you can't see through. Make two cans with each item. See whether your brothers and sisters and friends can match them up in pairs. Then see whether they can guess what's in the cans.

Teacher Information

This is a preschool or early-grade activity, although older students might enjoy it as well. Older children might be interested in constructing the "shakers" for younger groups. Materials used in the film canisters to make noise could include: dried rice, beans, or peas; marbles; BBs; gravel; sand; bits of plastic foam; puffed rice; or any other small objects found around the home or school. Be sure the canisters are prepared with approximately the same amount of material in each canister. Any small, opaque containers with lids can be substituted for the film canisters, of course. The containers need to be the same size and made of the same material.

The object of this activity is not to guess exactly what item is in the container, although that is a fun challenge, but to practice skills of observation by listening and skills of verbal expression. The risk in focusing on identifying the exact object is that students might do a good job of describing that the object is flat, thin, hard, and about 2 or 3 cm wide, then feel that they were wrong because they guessed the item to be a quarter, only to discover that it was a metal washer. If the descriptors are accurate, the response should be considered correct, because that is the objective of the activity.

An interesting adaptation is to prepare pairs of identical rattle cans, two with marbles, two with toothpicks, and so on. Number the cans randomly, but have a numbered list of the contents to provide a key if needed. Make several pairs, and have students match the pairs. Again, the objective is more to match sounds than to be fully accurate on the exact items contained in the cans.

Consider trying this also to match people up at a party or other event, to get them to become acquainted with each other in the group. Give each person in the group one rattle can, and instruct each to find his or her partner by finding a matching sound.

Integrating

Music

Science Process Skills

Observing, inferring, classifying, communicating, comparing and contrasting, identifying and controlling variables

How Can You Make Music with Fish Line?

(Take home and do with family and friends.)

Materials Needed

- 50 cm (20 in.) of monofilament fish line with a wooden dowel 10 cm (4 in.) long attached to one end
- Additional dowel
- Shoe box with lid
- Toothpick
- Pencil

Procedure

1. Put the shoe box on the edge of a table or desk.
2. Remove the lid, make a hole in one end of the box with your pencil, thread the fish line through the hole, and secure it with the toothpick, as in Figure 4.4–1.

Figure 4.4–1. Open Shoe Box

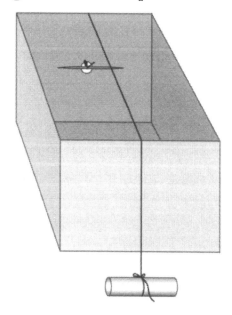

Hands-On Physical Science Activities

Figure 4.4–2. Box with Lid

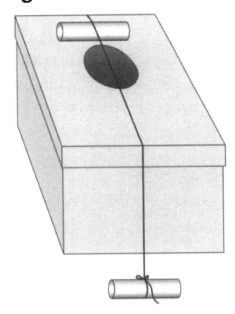

3. Put the lid on the box and stretch the fish line lengthwise across the top of the box. Put the additional dowel under the fish line near one end of the lid, as in Figure 4.4-2.

4. Slowly pull down on the dowel and pluck the fish line. What do you hear? What do you see?

5. As you pluck the line, pull down on the dowel to stretch it tighter. Watch and listen. What do you see? What do you hear? Try to play a simple tune. What might happen if you cut a hole in the lid of the shoe box? Try it.

For Problem Solvers

Experiment with different materials for the string on your homemade instrument. Try different types and weights of fish line, string, fine wire, or whatever is available. Which one can you get the highest notes with? Which one can you get the lowest notes with? Why do you think they are different?

Teacher Information

This activity should help children discover the relationship between the rate (speed) of vibration and the pitch (high to low) of sound. You may want to try this activity with rubber bands. The hole cut in the lid will increase the resonance. Resonance is a way of increasing the intensity of

Sound

a sound by causing one object (vibrating fish line) to create a sympathetic vibration of about the same frequency in another object (the walls of the shoe box).

The dowel on the top of the lid serves as a bridge to keep the string elevated enough to vibrate freely; stringed instruments use the same principles. A ukulele, guitar, violin, cello, or viola could be used for comparison.

Integrating

Music

Science Process Skills

Observing, classifying, measuring, predicting, communicating, comparing and contrasting, identifying and controlling variables, experimenting

Hands-On Physical Science Activities

How Much Noise Can You Make with a Paper Cup?

(Take home and do with family and friends.)

Materials Needed

- Plastic cup (or paper)
- String, about 30 cm (12 in.) long
- Half of a toothpick

Procedure

1. Make a small hole in the bottom of the cup, at the center. You can use the toothpick or your pencil to make the hole.
2. Insert the end of the string through the hole in the cup.
3. Tie the string around the half-toothpick, so the toothpick will be on the outside of the cup. (See Figure 4.5–1.)

Figure 4.5–1. Cup, String, and Toothpick

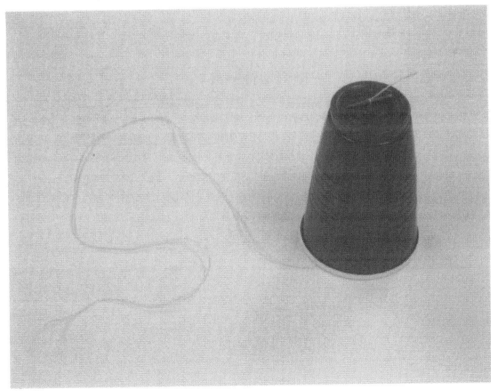

4. Hold the cup in one hand. Squeeze the string with the thumb and index finger of the other hand and pull, holding the string tight, but allowing it to slip through your fingers.

5. Did you hear anything? If not, squeeze the string a little tighter next time.

6. Wet the string with water and do it again.

7. What happened? Explain, and have fun!

For Problem Solvers

Do this project as a Science Investigation. Obtain a blank copy of the "Science Investigation Journaling Notes" from your teacher. Write your name, the date, and your question at the top. Plan your investigation through item 5 (Procedure) and have it approved by your teacher. Complete the Journaling Notes as you perform your investigation. Share your project with your group, and submit your Journaling Notes to your teacher if requested.

Experiment with different materials. The variables are the cup, the string, the water, and how tightly you held the string with your fingers. Change one variable at a time and see what differences you can make. Predict what difference each change will make before you try it.

If you can find a small feather or two and some rolly eyes, you could decorate this noisemaker and call it the yellow chicken, the red hen, or whatever you decide. You could fashion a beak out of a clothespin or some other suitable object.

Why does this thing make sound? What is the role of the cup?

Teacher Information

This activity is very noisy, so you might want to save it for the end of the day (or perhaps just before your students go to another teacher!). You might even want to do it outdoors. Students will enjoy it, though, and they will learn as they try different variables to make their yellow chicken, or their red hen, or whatever they decide to call the noisy thing. The friction between string and fingers causes the string to vibrate, and the sound is amplified by the cup.

Integrating

Music

Science Process Skills

Observing, classifying, measuring, predicting, communicating, comparing and contrasting, identifying and controlling variables, experimenting

How Can You Make Bottled Music?

(Take home and do with family and friends.)

Materials Needed

- Eight glass soda bottles of the same size and shape
- Water
- Paper slips numbered 1 to 8
- Pencil

Procedure

1. Pour water to different levels in the bottles.
2. Blow gently across the tops of the bottles until a sound is produced for each one. Arrange the bottles in a row according to the pitch of the sound from low to high.
3. You may want to add to or remove water from the bottles to make a musical scale. Under each bottle put a slip of paper (numbered from 1 to 8 for the eight bottles).
4. Try to play a simple tune by blowing across the tops of the bottles. Can you decide what is vibrating to make the sound?
5. Use a pencil to tap the side of each bottle near the top. What happened? What is vibrating to make the sound? What can you say about this?

For Problem Solvers

Try to tune the bottles to the piano. Can you make one of the bottles produce the same pitch as one of the piano keys? Can you match the piano with one full octave of sounds from the bottles? One octave is eight white keys in a row, ignoring the black keys between them. If you are using bottles with lids, put the lids on overnight and see whether the sounds still match the next morning.

Teacher Information

To do this activity, children may need to be reminded of the musical meaning of the term *pitch*.

In steps 1 through 4, blowing across the bottle causes the air to vibrate. This is the way pipe organs and musical wind instruments produce sound. A longer column of air will cause a slower vibration and a lower pitch. When the bottles are struck in step 5, it is the glass that vibrates to produce the sound. Water will slow the rate of vibration of the glass. Therefore, the greater the amount of water, the more slowly the glass vibrates, and the lower the pitch.

Remember, water expands and contracts according to its temperature. Water also evaporates. Both of these factors may make the bottles change pitch if they are kept for later use.

Integrating

Music

Science Process Skills

Observing, classifying, measuring, predicting, communicating, comparing and contrasting, identifying and controlling variables, experimenting

How Can You Make Music with Tubes?

(Take home and do with family and friends.)

Materials Needed

- Drinking straws
- Garden hose 1 m (1 yd.) long
- Scissors
- Mouthpiece from a bugle, trumpet, or trombone

Procedure

1. Cut one end of a drinking straw to a point, as in Figure 4.7–1. Moisten the cut end and put it between your lips. Blow gently around the straw. Cut pieces from the end of the straw as it is being played. What happens? What can you say about this?

2. Place a mouthpiece in a garden hose. Blow into the mouthpiece to see whether you can make a sound. Change the shape of the hose. What happens to the pitch of the sound?

3. Try it without the mouthpiece. Can you make the same sounds?

For Problem Solvers

Making musical sounds with the soda straw and with the garden hose probably gave you lots of new ideas for making still more musical sounds. Try your ideas. Get several different kinds and sizes of soda straws and do whatever you can do with them to make new sounds.

Put a skinny straw into a fat straw and slide them in and out as you play notes. What happened?

Cut one or more holes along the top of a straw and make more new sounds. Can you play it like a flute?

Figure 4.7–1. Straw with Cut End

Teacher Information

With practice, the students will be able to make the cut end of the straw vibrate to produce sound. This is similar to a clarinet or oboe. Paper straws work better than plastic because the plastic does not compress as easily to form a reed. Even so, plastic straws are very usable for this activity.

When the group uses the garden hose, a child who plays the trumpet, trombone, or bugle may be able to demonstrate and help others learn to play. Changing the shape of the hose (straight, coiled, and so forth) will not vary the pitch; however, cutting a length off either the straw or the hose will shorten the vibrating column of air and raise the pitch.

Integrating

Music

Science Process Skills

Observing, classifying, measuring, predicting, communicating, comparing and contrasting, identifying and controlling variables, experimenting

How Can You Make a Kazoo with a Comb?

(Take home and do with family and friends.)

Materials Needed
- Combs of various sizes
- Kazoos
- Tissue paper 10 cm by 20 cm (4 in. by 8 in.)

Procedure
1. Fold the tissue paper over the comb, letting it hang down each side.
2. Hum into the paper-wrapped comb. What happened? What can you say about this?
3. Hum a tune on the kazoo. How does it make a sound?
4. Raise your head so you are looking at the ceiling of the room. Hold your fingers on the front of your neck. Hum a tune. What happened? What can you say about this?

Teacher Information

Sounds are produced by vibrations. With the tissue paper and comb, sound is produced by the vocal cords, which in turn cause the paper over the comb to vibrate and alter the sound. The same occurs when a kazoo is used.

This activity can provide a review of the way sound is produced through vibration and how sounds can be altered through the vibration of another object.

Kazoos may be purchased in novelty stores that carry party noisemakers. Some music stores also carry them.

Integrating
Music

Science Process Skills
Observing, measuring, communicating, identifying and controlling variables, experimenting

What Is a Triple-T Kazoo?

(Take home and do with family and friends.)

Materials Needed

- Toilet-tissue tube
- A square of waxed paper (about twice the diameter of the tube)
- Rubber band
- Paper punch

Procedure

1. Punch a hole in one end of the toilet-tissue tube. Reach as far into the tube as you can with the paper punch.

2. Wrap the waxed paper over the other end (opposite the punched hole) and secure the waxed paper with the rubber band. (See Figure 4.9–1.)

Figure 4.9–1. Triple-T Kazoo

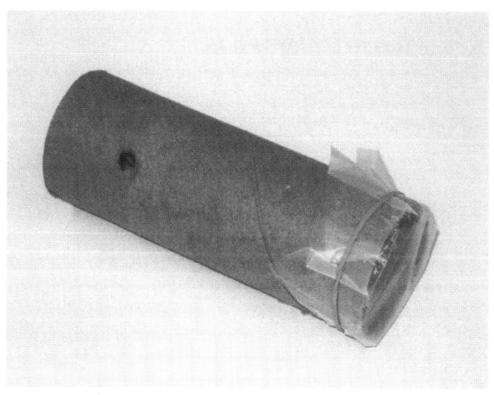

Hands-On Physical Science Activities

3. Hum into the open end of the tube.

4. Bring your kazoo to school and make music with others who are making kazoos and with other instruments of all kinds that have been made by the class.

5. Color your kazoo. Be creative, and make it just the way you want it to be.

6. Why do we call it the "Triple-T" kazoo? When you think you know, tell your teacher, but keep it a secret from those who are still trying to decide.

Teacher Information

The sounds, and change in pitch, are produced by the voice, but the kazoo gives it an interesting sound. Humming into the open end of the tube causes the paper on the other end to vibrate at the same frequency as the voice, thus producing the same pitch as the one that is hummed. Children will enjoy adding the Triple-T kazoo to their collection of homemade musical instruments.

Why do we call it the Triple-T kazoo? Because it's made from a Toilet-Tissue Tube, of course!

Integrating

Music, art

Science Process Skills

Observing, experimenting

How Can You See Sound?

Materials Needed

- Cardboard oatmeal drum or medium-sized tin can per small group
- Puffed rice or salt
- Heavy rubber bands (or string)
- Large balloons or sheets of rubber

Procedure

1. Remove both ends from a can or cardboard container (be careful of sharp edges).
2. Stretch a balloon over one end of the can and secure it with the rubber band. You now have a simple drum, similar to the one in Figure 4.10–1.

Figure 4.10–1. Tin-Can Drum

3. Sprinkle puffed rice on the drum head. Tap the drum head and observe what happens. Hit it harder. What happened? What can you say about this?

4. Sprinkle puffed rice or salt on the drum head. Keep the drum level and shout into the other end. Have a friend observe the results, then trade places.

5. What happened? Discuss your observations?

For Problem Solvers

Do this project as a Science Investigation. Obtain a blank copy of the "Science Investigation Journaling Notes" from your teacher. Write your name, the date, and your question at the top. Plan your investigation through item 5 (Procedure) and have it approved by your teacher. Complete the Journaling Notes as you perform your investigation. Share your project with your group, and submit your Journaling Notes to your teacher if requested.

Put some salt on the drum head. Hold it over the speaker of a stereo while you play music on the stereo. Turn the volume up and down. What is happening with the salt? Why? Do you see any difference when the volume changes? Do you see any difference with high notes and bass notes?

Teacher Information

When the drum head is tapped, the salt will form a pattern caused by the vibration. The pattern will change as the drum is hit harder.

Shouting into the can will cause the rubber diaphragm to vibrate (be sure it is tightly stretched). Pitch and loudness will change the pattern of the salt or puffed rice.

Loud musical instruments will also cause the patterns to change. Remember, the balloon must be stretched tightly across the can.

Integrating

Music

Science Process Skills

Observing, measuring, predicting, communicating, comparing and contrasting, identifying and controlling variables, experimenting

What Can Water Teach Us About Sound?

(Whole-group activity)

Materials Needed

- Box of dominoes
- Pan of water
- Small rocks
- Drawing paper
- Crayons

Procedure

1. Stand the dominoes on end on a solid surface approximately 3 cm (1 in.) apart.

2. Tip the first domino forward so it hits the one next to it. What happened?

3. Matter is made up of tiny particles called molecules that react very much as the dominoes did when the first one was disturbed. With sound, energy in the form of vibration is transferred from one molecule to another.

4. Drop a small rock into a pan of water. Observe what happens to the water. When molecules bump against one another, they transfer energy to all the other molecules around them, causing ripples (little waves) of water to travel in all directions, much the same as sound waves do.

5. Can you draw a picture of the way you think sound travels in air? Try it, and show your picture to your teacher.

Teacher Information

The use of dominoes will help children see how energy is transferred from one object to another. The pan of water should show how the energy is transferred in all directions (in this case, in ripples). If a large pan is used, the children may observe that the ripples bounce off solid objects and reverse direction. Echoes are caused by sound waves traveling out and bouncing back in waves.

Integrating

Music

Science Process Skills

Observing, measuring, predicting, communicating, identifying and controlling variables, experimenting

How Many Ways Can You Make a Telephone?

(Take home and do with family and friends.)

Materials Needed

- Two paper cups
- Toothpicks
- Cotton thread, about 4 m (4 yd) long
- Partners

Procedure

1. Use the toothpick or your pencil to punch a small hole in the center of the bottom of each cup.
2. Push one end of the thread through the hole of each cup.
3. Tie each end of the thread to the toothpick so the thread cannot pull out of the hole in the cup. (See Figure 4.12–1.)
4. Keep the thread tight and be sure it doesn't touch anything.
5. Put the cup to your ear and have a friend talk into his or her cup. Now you talk and have your friend listen. Now whisper.
6. What happened? What happens when you touch the thread while you're listening with the cup to your ear? Explain why you think this happens. Make a set of telephones and show them to your family.

Figure 4.12–1. Paper Cup with Toothpick and Thread

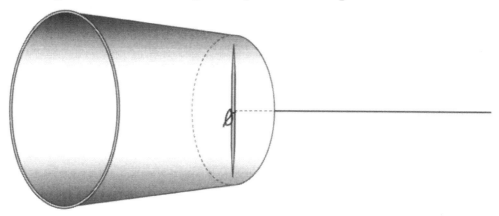

For Problem Solvers

Make a "party line" by crossing the lines from two sets of telephones over each other. Three people can then listen while one person talks. See whether you can include a third set of telephones.

Try replacing the string with fishing line, wire, dental floss, or heavier and lighter string. Try empty cans in place of the cups. Try larger or smaller cups. Each time you try something new, talk about it and predict what the difference will be before you test the change. What other investigations can you think of for the homemade telephone?

Teacher Information

The telephone works in a very simple way. Sound waves cause the bottom of the first cup to vibrate. These vibrations, in turn, cause the thread to vibrate. The vibrating thread causes the bottom of the other cup and the air inside it to vibrate. The sounds the students hear are a result of these vibrations; the air in the receiving cups strike their eardrums in nearly the same way it struck the bottom of the first cups as their partners spoke into them.

Use heavy cotton thread. Polyester is easier to find, but it tangles easily. Dental floss is an excellent substitute, but more expensive.

Integrating

Music

Science Process Skills

Observing, measuring, predicting, communicating, identifying and controlling variables, experimenting

ACTIVITY 4.13

How Well Does Sound Travel Through Wood?

(Take home and do with family and friends.)

Materials Needed

- Table (or desk)
- Partners

Procedure

1. Have a partner tap an object on the table or desk loudly enough for you to hear.
2. Put your ear on the desk top and have your partner tap again.
3. What happened?
4. What can you say about this?
5. What can you say about sound traveling through solid objects? Can you think of a reason for this?
6. What does the statement "Keep your ear to the ground" mean? Where do you think it began?

Teacher Information

Sound travels better through solid objects because the molecules are more tightly packed. They bump against one another more frequently and transmit the vibrations more efficiently. Sound will travel a *greater distance* in solids for the same reason. The exception, of course, is specially designed acoustic materials that appear to be solid but are designed with many spaces, each of which has its own walls that reflect vibrations in different directions, having the effect of softening the sound.

The tapping on the desk will be heard more clearly when the ear is against the desk.

Native Americans used this principle, literally keeping their ears to the ground, to hear sounds at great distances. Buffalo herds and horses' hooves could be heard before they were seen. "Keep your ear to the ground" has come to mean "listen carefully."

Integrating

Social studies

Science Process Skills

Observing, predicting, communicating, identifying and controlling variables

244

Hands-On Physical Science Activities

From How Far Away Can You Hear a Clock Tick?

(Take home and do with family and friends.)

Materials Needed

- Ticking clock
- Foot ruler
- Meter stick (or yardstick)
- String
- Paper-cup telephone
- Partners

Procedure

1. Listen to the clock tick. Move it as far from your ear as you can and still hear the ticking. Have a partner measure the distance.

2. Put one end of the foot ruler to your ear and the clock at the other end, touching the ruler. What happened?

3. Repeat step 2 using a meter stick. Substitute string for the meter stick. What happened?

4. Put the clock against one end of your paper-cup telephone and listen on the other end. What can you say about this?

5. The sticks should not be held firmly or clutched in your hand. Why?

For Problem Solvers

Find a long board, preferably a 1- by 2-inch that is 8 feet long. Be sure you have your measurement recorded from step 1 above, which indicates how far away you could hear the clock tick through the air. Put the board against your ear and have a friend put the clock against the board. Move the clock farther away until you can no longer hear it. Measure this distance and write it down. Could you hear the clock from any farther away through the board than you could through the air? Does sound travel better through air or through solids?

Teacher Information

Thin pieces of wood such as lath of different lengths may be substituted for the foot ruler and meter sticks. The objects should not be held firmly, as the hand will absorb the vibrations and muffle the sound. The same phenomenon occurs when the thread on the paper-cup telephone is touched.

For your problem solvers, 1- by 2-inch lumber is often called firring, and it is easy to find at lumber stores or home improvement stores. Any long, thin board will do just fine.

Integrating

Math

Science Process Skills

Observing, measuring, predicting, communicating, comparing and contrasting, identifying and controlling variables, experimenting

Hands-On Physical Science Activities

How Can You Make a Coat Hanger Sing?

(Take home and do with family and friends.)

Materials Needed

- Metal coat hanger
- Two heavy cotton strings 50 cm (20 in.) in length
- Copies of the "Science Investigation Journaling Notes" for this activity for all students

Procedure

1. As you complete this activity, you will keep a record of what you do, just as scientists do. Use a copy of the form "Science Investigation Journaling Notes" and write the information that is called for, including your name and the date.

2. For this activity, you will learn how to get interesting sounds from a coat hanger. For item 1, the question is provided for you on the form.

3. Item 2 asks for what you already know about the question. If you have some ideas about how sounds are made, write your ideas.

4. For item 3, write a statement of how you might be able to make a coat hanger sing, based on what you already know, and that will be your hypothesis.

5. Now continue with the following instructions. Complete your Journaling Notes as you go. Steps 6 through 10 below will help you with the information you need to write on the form for items 4 and 5.

6. Tie the strings to the wide ends of the hanger, as shown in Figure 4.15-1.

7. Hold the ends of the strings and hit the hanger against a solid object, such as a desk or table. Listen to the sound it makes.

8. Wrap the ends of the string around each of your index fingers. Put your fingers in your ears and tap the hanger on the solid object again.

9. Compare the first sound with the one you just heard.

Sound

Figure 4.15–1. Coat Hanger with Two Strings

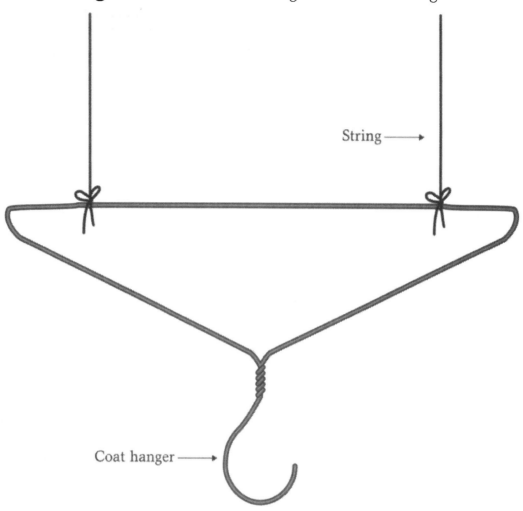

String →

Coat hanger →

10. What caused the sound? Discuss your ideas with your family.

11. Complete your "Science Investigation Journaling Notes." Are you ready to explain why how these sounds come from a coat hanger? Discuss it with others who did this activity.

12. Can you see a relationship between this activity and the one with the clock?

For Problem Solvers

Do this project as a Science Investigation. Obtain another blank copy of the "Science Investigation Journaling Notes." Write your name, the date, and your question at the top. Plan your investigation through step 5 (Procedure)

Hands-On Physical Science Activities

and have it approved by your teacher. Complete the Journaling Notes as you perform your investigation. Share your project with your group and submit your Journaling Notes to your teacher if requested. Were you surprised about the sound you heard from the coat hanger? Investigate this idea further. What variables could you change? Can you substitute something else for the coat hanger, such as a spoon, a piece of wire, another kind of coat hanger, or something else? Could you use a lighter or heavier string, or one of a different length, or a thin wire, or something else? Try tapping it against various objects and surfaces as well.

How does a doctor's stethoscope work? Do some research and find out. Do you see any relationship between the coat hanger activity and the stethoscope? Explain.

Did you know that mechanics also use stethoscopes? The mechanic's stethoscope usually has a long, narrow probe instead of a broad pad. Think of ways you think a stethoscope might be helpful to a mechanic. Now go back to the research and find out if you were right.

Teacher Information

When struck without the fingers in the ears, the hanger will sound flat and metallic. When the fingers are placed in the ears, the sound will be a loud gong because sound travels better through solids (string) than through air.

When an unwanted sound is coming from the engine of your car, the sound resonates through other parts, making it very difficult to pinpoint the source of the sound. Your mechanic might even recognize it as the sound of a worn bearing, yet be unable to determine accurately whether it's coming from the alternator, the power steering pump, the air conditioning compressor, or just where. By checking these potential trouble points with the probe of the mechanic's stethoscope, the sound is remarkably more vivid when the probe contacts the trouble spot.

Integrating
Music

Science Process Skills
Observing, measuring, predicting, communicating, comparing and contrasting, identifying and controlling variables, experimenting

Science Investigation
Journaling Notes for Activity 4.15

1. Question: *How can you make a coat hanger sing?*
2. What we already know:

_____

_____

3. Hypothesis:

_____

_____

4. Materials needed:

_____

_____

5. Procedure:

_____

_____

6. Observations/New information:

_____

_____

7. Conclusion:

_____

_____

How Fast Does Sound Travel?

Materials Needed

- Drum, cymbals, large metal lid or something else that will make a loud sound when visibly struck
- Stick to strike object

Procedure

1. Take your drum or other object out on the school grounds. Ask other members of the class to go with you.
2. Move about 100 meters (or about 100 yards) or more away from the other students.
3. Strike the object several times so the others can see the movement of your arm and hear the sound.
4. Remember, when you see an object move at a distance, you are seeing reflected light. When you hear the sound, you are hearing sound vibrations.
5. Tell what you observed. What can you say about the speed of light and the speed of sound?

For Problem Solvers

Do this project as a Science Investigation. Obtain a blank copy of the "Science Investigation Journaling Notes" from your teacher. Write your name, the date, and your question at the top. Plan your investigation through item 5 (Procedure) and have it approved by your teacher. Complete the Journaling Notes as you perform your investigation. Share your project with your group, and submit your Journaling Notes to your teacher if requested.

Using a stopwatch, have someone strike a metal post with a hammer from 100 meters away. Figure out a way to measure the distance. A greater distance (200 or 300 m) would be even better, but be sure you know how far it is. Start the stopwatch when you see the hammer strike the pole, and stop it when you hear the hammer strike the pole. Check the time several times for accuracy. Figure out how far sound travels in one second. How long does it take sound to travel one kilometer? Translate that to miles. How long does it take sound to travel one mile?

Teacher Information

Even at the short distance of 100 meters, it will be possible to see the child strike the drum before the sound is heard. Children who have attended athletic events in a large stadium may have noticed that movements by athletes or bands that create sound are seen before the sounds are heard. Airplanes are sometimes difficult to locate in the sky by their sound because, by the time the sound arrives, the plane has moved to a different position.

Light travels very rapidly, at about 300,000 km (186,000 mi.) per second. By comparison, sound is a slowpoke, moving at about 330 m (360 yd.) per second at sea level, and at freezing point. This is about 1,190 km (740 mi.) per hour.

The speed of sound is not affected by loudness or frequency. It is affected slightly, however, by temperature and by density of the air (thus, slightly faster at sea level than at a higher altitude where the air is less dense). As temperature rises, air molecules move faster and therefore bump into one another more frequently. The rate of increase in speed is about 0.6 meters per second for every Celsius degree. Sound travels about four times faster in water (fifteen times faster in steel) than it does in air.

Integrating

Math, language arts

Science Process Skills

Observing, inferring, measuring, predicting, communicating, using space-time relationships, formulating hypotheses, identifying and controlling variables, experimenting

What Is a Tuning Fork?

Materials Needed

- Tuning fork
- Small dish of water
- Soft rubber mallet or rubber heel of a shoe

Procedure

1. Take turns holding the handle of the tuning fork in one hand and strike it with a rubber mallet or the heel of a shoe. (*Caution:* You shouldn't hit the tuning fork with a hard object.)
2. Bring the double end near your ear. What happened?
3. Strike the tuning fork again. Touch the double end with your finger. How did it feel?
4. This time, after striking the tuning fork, lower it slowly into the dish of water. What happened?
5. Strike the tuning fork again. Gently touch the handle to a hard surface such as a table or desk. Discuss your observations.

For Problem Solvers

You noticed what happened when you touched the base of the tuning fork to the table. Experiment with the tuning fork to see whether you can find other ways to make it louder. Try touching it to a box, the body of a guitar or violin, and other things. What materials and what shapes seem to have the greatest effect?

Teacher Information

Tuning forks may be obtained from music stores and science-supply houses. Children who play stringed musical instruments may have tuning forks for use in tuning their instruments. Doctors sometimes use tuning forks for general hearing tests. Each tuning fork is tuned to a certain pitch. The pitch depends on the thickness and length, and the material of which it is made.

Most tuning forks vibrate so rapidly that it is difficult to detect movement by looking at them. Your students will learn that when the tines of a vibrating fork are lowered slowly into a dish of water, the water will splash, demonstrating that vibration is occurring. When the handle of a vibrating fork is placed on a table or desk, the sound will be amplified. If the handle of the tuning fork is touched to a large paper cup, a shoebox, or the body of a stringed instrument, the sound will be amplified.

Integrating

Music

Science Process Skills

Observing, inferring, predicting, communicating, comparing and contrasting, identifying and controlling variables, experimenting

How Can You Make a Goblet Sing?

(Teacher demonstration)

Materials Needed

- Four to six good-quality glass goblets
- Water
- Vinegar

Procedure

1. Check the goblets carefully to be certain they have no cracked or chipped edges.
2. Add different amounts of water to each goblet (no more than half full). Put a few drops of vinegar in the water.
3. Firmly hold the goblet by the base with one hand. Moisten the fingers of your other hand with the vinegar water and rotate your fingers lightly around the rim of the goblet. Ask students what happened? Why do they think it does that?
4. Try the other goblets. Ask the students to describe what is happening.

For Problem Solvers

Do this project as a Science Investigation. Obtain a blank copy of the "Science Investigation Journaling Notes" from your teacher. Write your name, the date, and your question at the top. Plan your investigation through item 5 (Procedure) and have it approved by your teacher. Complete the Journaling Notes as you perform your investigation. Share your project with your group, and submit your Journaling Notes to your teacher if requested.

Take two goblets that are just alike. Be sure they are clean, dry, and that they have no chips or cracks around the rim. Place them about 30 cm (12 in.) apart. Hold one goblet steady while you moisten your finger with vinegar water and make this goblet sing. While it is producing a loud tone, grasp it firmly to stop the tone and immediately listen carefully to the other goblet. What do you hear? If you hear nothing from the other goblet, try it again.

Sound

Lift the lid of a piano and make a steady singing tone into the piano with your voice, while someone else holds the sustain pedal down. Stop the tone and listen. What do you hear?

Place two guitars face to face, just a short distance apart. Be sure they have been tuned alike. Pluck one string of one guitar, then stop the vibration of that string by placing your hand over it and listen to the other guitar. What do you hear?

See what you can learn about sympathetic vibrations. Use your encyclopedias, the Internet, or other references available to you.

Teacher Information

Great care should be exercised in performing this activity. The goblets must be of high quality and completely free of rough edges. When this activity is properly performed, the moist fingers will cause the glass to vibrate and produce a beautiful, clear tone. The combination of water and vinegar seems to produce just enough lubricant and friction to make the demonstration easier.

Your problem solvers will experiment with sympathetic vibrations. This means that the vibrations of one goblet will travel through the air and cause an identical goblet to vibrate at the same frequency, producing the same tone. In order for this to occur, the condition of each glass must be almost exactly the same. Both should be dry, empty, and at the same temperature. Their physical appearance should be the same.

Sympathetic vibration can also be experienced by humming a steady tone into a piano while holding the sustain pedal down. The vibrations of the voice will cause strings tuned to the same pitch in the piano to vibrate.

Integrating

Music

Science Process Skills

Observing, inferring, classifying, predicting, communicating, formulating hypotheses, identifying and controlling variables, experimenting, researching

How Can You Play a Phonograph Record with a Sewing Needle?

(Teacher-directed activity)

Materials Needed

- Paper cups (or plastic)
- Thin sewing needles
- Thimbles
- Masking tape
- Sheets of construction paper
- Old phonograph records
- Turntable

Procedure

1. Thomas Edison, a famous inventor, invented a talking machine that he called a phonograph. At first a wax cylinder was used. It was later improved by making a round, hard disk.

2. Very carefully push a sewing needle through the bottom edge of a cup at an angle, as seen in Figure 4.19–1.

3. Put an old phonograph record (it might get damaged) on a revolving turntable and hold the cup by two fingers while lightly touching the point of the needle to the grooves in the record, as shown in Figure 4.19–2. The needle should be slanted away so that it does not dig into the record but instead rides smoothly in the groove. What happened?

4. Form a sheet of construction paper into a cone held together with masking tape. Insert a sewing needle through the narrow part of the cone, as in Figure 4.19–3.

5. Lightly touch the needle to the grooves of the record spinning on the turntable. What happened? Can you explain why?

Sound

Figure 4.19–1. Paper Cup with Needle at an Angle

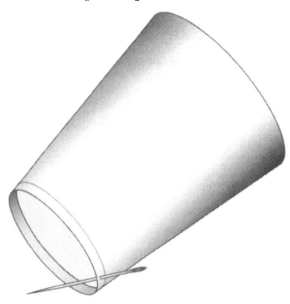

Figure 4.19–2. Paper Cup Holding Needle over Record

Hands-On Physical Science Activities

Figure 4.19–3. Needle Inserted Through Cone

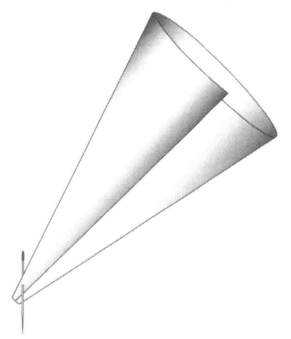

For Problem Solvers

Do this project as a Science Investigation. Obtain a blank copy of the "Science Investigation Journaling Notes" from your teacher. Write your name, the date, and your question at the top. Plan your investigation through item 5 (Procedure) and have it approved by your teacher. Complete the Journaling Notes as you perform your investigation. Share your project with your group, and submit your Journaling Notes to your teacher if requested.

Identify variables that you think might affect the production of sound in the above activity and experiment with those variables. Try various needles or other objects that you think might substitute for the needle. You could try many different ideas in place of the paper cup. Can you think of another way to turn the record, so you don't need the motorized turntable? Explore, and have fun learning about sound in a creative way.

You already know that sounds come from vibrations. Examine the record carefully with a hand lens. Use a microscope also if you have one. Can you tell what causes the needle to vibrate? Can you see anything that would cause it to change its vibrating pattern to create different sounds? Does a needle work any better for this than a straight pin? Examine the points and examine the grooves again. Look at the points of each under the microscope or hand lens while they are resting in a groove of the record. Which would you expect to work better? Why?

Teacher Information

The grooves in the record cause the needle to vibrate, which transfers to the bottom of the paper cup. The shape of the cup concentrates and amplifies the sound, making it possible to hear the sound reproduced from the grooves of the record. The paper cone works in the same way, except the sound may be louder due to the larger size of the cone and the better vibration that is possible through thinner material.

Perhaps when you were a child you thought there was something magical about a phonograph, but it is hard to see anything magical about a paper cup or a sewing needle. Yet these simple materials produce music and voice from a phonograph record. Those who do the activities for problem solvers will find that the grooves in the record are bumpy. These bumps cause the needle to vibrate, which in turn causes the cup or paper cone to vibrate, and the cup or paper cone amplifies the sound. They should also notice that the bumps are not evenly spaced. The closer together the bumps are, the faster the needle vibrates, making the pitch go up, with the inverse being true where the bumps are farther apart.

Caution: There are obvious safety concerns when using a needle. Young children will especially need help in inserting the needles and rolling the cones.

Integrating

Music

Science Process Skills

Observing, inferring, communicating, comparing and contrasting, formulating hypotheses, identifying and controlling variables, experimenting

How Can Sound Be Directed?

(Take home and do with family and friends.)

Materials Needed

- Two megaphones (can be large paper cones)

Procedure

1. Go with a partner to the playground. Move far enough apart that you need to shout to hear each other.
2. Speak to each other without using the megaphones.
3. Hold a megaphone up to your mouth, pointed in the direction of your partner, and speak with about the same voice volume as you did without the megaphone.
4. Take turns speaking to each other with and without the megaphones.
5. Each of you turn your megaphone in the opposite direction from your partner, and continue speaking.
6. Move farther apart and continue your conversation.
7. Get back together and discuss your observations. What differences did you notice in how easily you heard each other with and without using the megaphones? What difference did it make when you pointed the megaphone away, instead of toward, your partner?

For Problem Solvers

Do some investigating to see how many applications you can find for the megaphone. Who uses it? When? Why? Notice the shape of the end of a trumpet, where the sound leaves the trumpet. Is it shaped like a megaphone? Why? What other musical instruments can you find that use this shape?

Teacher Information

This activity is related to the activity with the paper-cup telephone. The megaphone collects, concentrates, and directs sound waves, as did the cups on the telephones. The children on the playground will discover that they can hear better when the megaphone is pointed toward them. We often cup our hands around our mouths when we shout, for the same reason. Many musical instruments implement the bell shape (cone shape).

Integrating

Music

Science Process Skills

Observing, inferring, classifying, measuring, predicting, communicating, comparing and contrasting, formulating hypotheses, identifying and controlling variables, experimenting, researching

How Can Sound Be Collected?

(Take home and do with family and friends.)

Materials Needed

- Two megaphones (can be large paper cones)

Procedure

1. Go with a partner to the playground. Move far enough apart that you need to shout to hear each other.
2. Speak to each other without using the megaphones.
3. Hold your megaphone with the small end up to your ear and the broad end pointed toward your partner while your partner speaks to you.
4. Take turns speaking to each other with and without the megaphones. When you are the listener, hold the small end of the megaphone up to your ear some of the time, and listen without the megaphone some of the time.
5. Move farther apart and continue your conversation.
6. Get back together and discuss your observations. What differences did you notice in how easily you heard each other with and without using the megaphone as a listening device?

Teacher Information

In this activity, incoming sound waves are reflected inward to the small end of the megaphone. This might be compared to the function of a funnel. When water is poured into the large end of the funnel it is "focused" into the small end. As sound waves enter the large end of a megaphone, they are focused toward the small end. Those with limited hearing often cup their hands behind their ears to produce the same effect. This activity is very closely related to the previous one, yet they focus on two different effects of the focusing of sound waves. The differences will be more clearly perceived by students if these are highlighted as separate activities.

Integrating

Music

Science Process Skills

Observing, inferring, classifying, measuring, predicting, communicating, comparing and contrasting, formulating hypotheses, identifying and controlling variables, experimenting, researching

Sound

What Are Sympathetic Vibrations?

Materials Needed

- Two identical tuning forks
- Two guitars, identically tuned

Procedure

1. Holding one of the tuning forks in each hand, strike one of them on the heal of your shoe, then immediately hold the two tuning forks side by side. The tuning forks should be near each other but must not touch each other.

2. Now touch the first tuning fork (the one that was struck) with your finger to stop the vibration.

3. Listen carefully to the second tuning fork. What do you hear?

4. Stand the two guitars on the floor, face-to-face. These also should be near each other, but they must not touch each other.

5. Pluck one string of one guitar, then touch it to stop its vibration and listen carefully to the other guitar.

6. What did you hear? Discuss your ideas with others who are involved with the activity.

Teacher Information

Each vibrating object has its own natural frequency. When two objects having the same natural frequency are placed near each other, the vibration of one will often transfer the energy through the air and the other object will vibrate. These are called *sympathetic vibrations*.

Fine crystal can respond to a sound from the voice of a singer if the pitch is the same as the natural frequency of the crystal. If the sound is loud and sustained at the natural frequency of the crystal, the crystal can actually shatter from intense sympathetic vibrations. In 1940 the newly

Hands-On Physical Science Activities

constructed Tacoma Narrows bridge, in the state of Washington, was destroyed by sympathetic vibrations that were started and sustained by a mild gale. The sympathetic vibrations increased in intensity, causing the bridge to swing and twist at its natural frequency until the bridge collapsed. The bridge is remembered as the "Galloping Gertie."

Integrating
Language arts, music

Science Process Skills
Observing, inferring, communicating

What Are Forced Vibrations?

Materials Needed

- Tuning forks
- Table (preferably solid wood construction)

Procedure

1. Strike the tuning fork on the heel of your shoe, then immediately place the stem of the tuning fork on the table. Listen carefully.
2. What did you hear? Discuss your observations and ideas with others who are involved with the activity.

Teacher Information

Sometimes when a vibrating object comes in contact with another solid object, the second object vibrates at the same frequency as the first object, even if the two objects have different natural frequencies. This is called *forced vibration*. The wooden table will likely vibrate at the natural frequency of the tuning fork in this activity, and with its large mass it will amplify the sound.

Integrating

Language arts

Science Process Skills

Observing, inferring, communicating

What Is the Doppler Effect?

(Teacher demonstration)

Materials Needed

- Tape recorder with blank audiotape
- Car and driver

Procedure

1. Stand with your students beside a road near the school.
2. Arrange for an adult to drive a car past the group. The driver should drive at a steady speed (but not too slowly) and begin honking the horn about half a block away and not let off the horn until the car is about half a block past the group.
3. Start recording before the horn begins to honk and continue to record until the horn stops honking.
4. Discuss student observations. What changes were noticed in the sound of the horn as the car went by?
5. The experience can be reviewed and discussed further by listening to the recording.

Teacher Information

If circumstances are not conducive to blasting a horn past the school, second-best is to pre-record the experience and use the recording for class discussion. The live experience is more meaningful, of course.

The faster the car is moving as it passes by, the more dramatic will be the change in pitch with the Doppler effect. If a sound source is moving rapidly as it approaches you, it pushes into the sound waves, having the effect of shortening the waves and raising the pitch. As the sound source retreats, it has the opposite effect and lowers the pitch. Figure 4.24–1 shows how the waves of water are compressed in front of a boat in forward motion and lengthened behind it. These waves, of course, are easily seen by an observer. Figure 4.24–2 illustrates how sound waves, which are not seen, are similarly compressed in front of the blasting horn of a car in forward motion and lengthened behind it.

Remember to make connections between the observation and related science concepts: Longer waves produce lower pitch and shorter waves produce higher pitch.

Sound

Figure 4.24–1. Boat Moving Forward Through Water

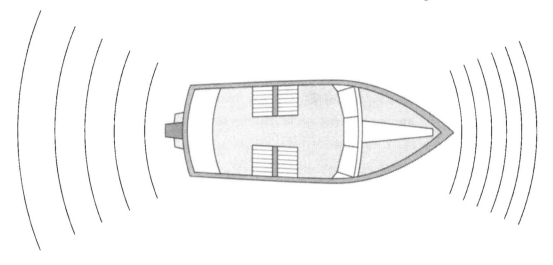

Figure 4.24–2. Car Moving Forward Through Air

Integrating
Language arts

Science Process Skills
Observing, inferring, communicating

What Is a Sonic Boom?

(Teacher-led discussion)

Materials Needed

- None

Procedure

Discuss any experiences students recall with a sonic boom. Add to the discussion with the following information.

Teacher Information

A common misconception with the sonic boom is that the "boom" occurs only at the moment that an aircraft accelerates past the speed of sound, or "breaks the sound barrier." The effect actually follows the aircraft as long as it is traveling at speeds in excess of the speed of sound.

Consider an automobile theoretically under continuous acceleration until it exceeds the speed of sound. Due to the Doppler effect, sound waves in front of the automobile would become shorter and shorter until the automobile reached the speed of sound. Then the sound waves would simply spread out to the side and behind the vehicle. Compressed sound waves and compressed air would actually become a barrier in front of the vehicle. Even with an accelerating boat, there comes a point at which the driver experiences a somewhat similar effect as the waves pile up in front of the boat. The driver has to increase power to push over the barrier of waves; then further acceleration is easier.

In order for an airplane to accelerate beyond the speed of sound, it must push through a barrier of compressed air. Doing so requires more power, and it produces a shock that can severely stress the structure of the plane. Once through this barrier, further acceleration is easier. The shock can be heard on the ground as a sonic boom, which then follows along with the aircraft in a cone-shaped pattern. The shock wave continues with the plane until the aircraft slows below the speed of sound. (See Figure 4.25–1.)

If possible, get a pilot or someone else knowledgeable in the field of supersonic speed to discuss with the class what the sonic boom is, its effect on people, and related flight regulations.

Figure 4.25–1. Cone-Shaped Sound Path

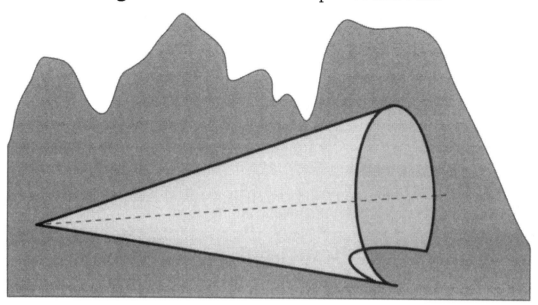

Integrating
Language arts

Science Process Skills
Observing, inferring, communicating

Can You Invent or Make a Musical Instrument?

(Teacher-assisted activity)

Materials Needed

- A variety of tubes, cans, rubber bands, and so on, that will produce sounds
- Figures 4.26–1, 4.26–2, and 4.26–3 to hand out to students

Procedure

1. Throughout history, people have made and played musical instruments. The only rule seems to have been that the sound an instrument made was pleasing to the person playing it. Hollow logs were probably the first drums; reeds, the first wind instruments; and tough stems or dried animal parts, such as tendons or intestines, the first stringed instruments. Today, much of our music is produced by electronics. But many of the old ways of producing music are still being used, and some old ways of producing music are being revived. Most of the instruments being used in symphony orchestras of today were invented hundreds and even thousands of years ago.

2. You can invent and play your own musical instrument. Look at the illustrations here. These ideas should help you begin. Your instrument does not have to be the same. Maybe you can invent a better one.

3. When you have finished making your musical instrument, find classmates who have made instruments and see whether you can learn to play a tune together.

Teacher Information

This activity can be used in conjunction with an art class or a music class. Encourage students to be creative as they make their instruments.

Integrating

Music, art

Science Process Skills

Observing, inferring, classifying, measuring, predicting, communicating, comparing and contrasting, using space-time relationships, formulating hypotheses, identifying and controlling variables, experimenting, researching

Figure 4.26–1. Homemade Percussion Instruments

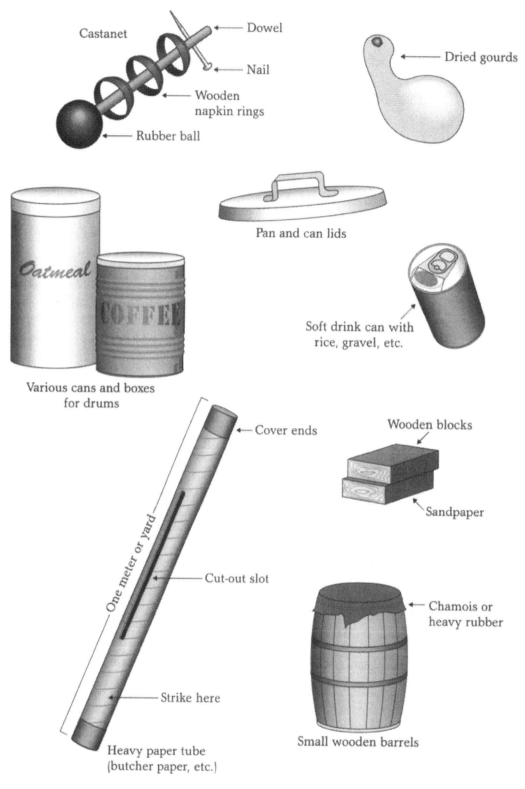

Castanet

Dowel

Nail

Wooden napkin rings

Rubber ball

Dried gourds

Pan and can lids

Various cans and boxes for drums

Soft drink can with rice, gravel, etc.

Cover ends

One meter or yard

Cut-out slot

Strike here

Heavy paper tube (butcher paper, etc.)

Wooden blocks

Sandpaper

Chamois or heavy rubber

Small wooden barrels

Figure 4.26–2. Homemade String Instruments

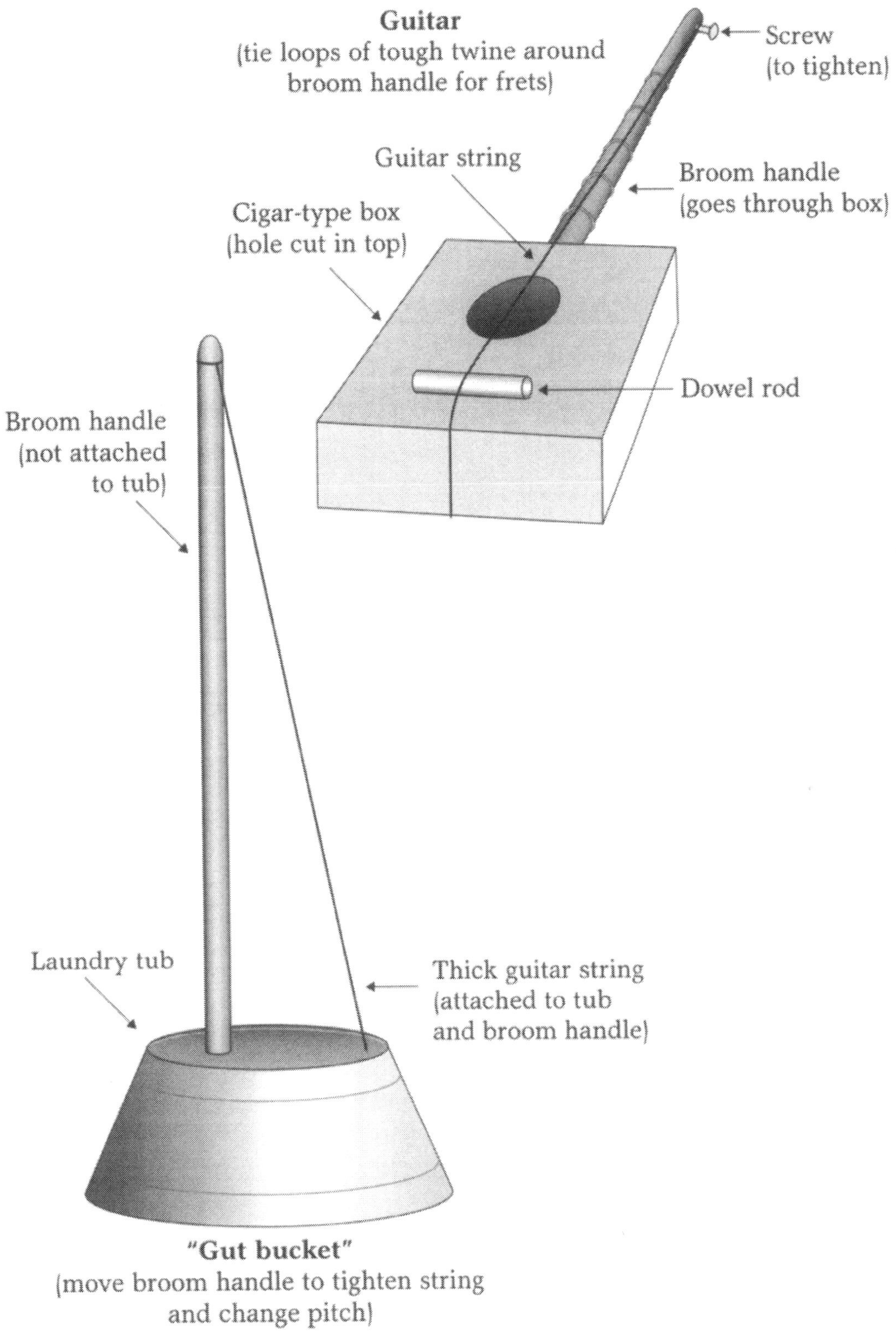

Guitar
(tie loops of tough twine around broom handle for frets)

Screw
(to tighten)

Guitar string

Broom handle
(goes through box)

Cigar-type box
(hole cut in top)

Dowel rod

Broom handle
(not attached
to tub)

Laundry tub

Thick guitar string
(attached to tub
and broom handle)

"Gut bucket"
(move broom handle to tighten string
and change pitch)

Figure 4.26–3. Homemade Wind Instruments

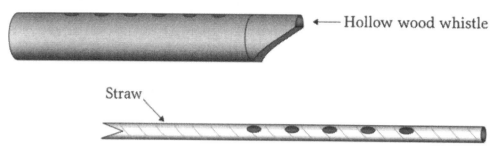

← Hollow wood whistle

Straw

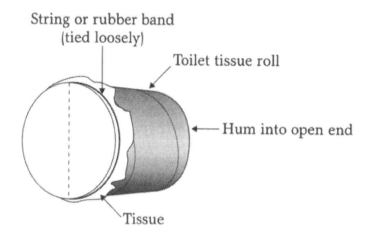

String or rubber band
(tied loosely)

Toilet tissue roll

← Hum into open end

Tissue

Hands-On Physical Science Activities

Can You Solve This Sound Word Search?

Try to find the following Sound terms in the grid below. They could appear in horizontal (left to right), vertical (up or down), or diagonal (upward or downward) position.

sound	vibration	amplify
music	noise	pitch
resonance	loud	soft
vocal cords	energy	tone
voice	hear	vibrate
molecules		

```
N  B  V  C  X  V  I  B  R  A  T  E
E  S  M  U  S  I  C  D  F  M  G  H
N  Y  U  I  O  B  P  V  L  P  K  J
E  H  T  M  R  R  E  O  O  L  W  Q
R  E  S  O  N  A  N  C  E  I  C  Z
G  A  M  L  N  T  B  A  V  F  C  X
Y  R  L  E  H  I  K  L  J  Y  H  E
R  E  E  C  W  O  S  C  S  D  F  G
T  S  T  U  Y  N  S  O  F  T  U  I
C  I  O  L  V  B  N  R  U  M  P  O
P  O  N  E  L  O  U  D  X  N  Z  A
K  N  E  S  J  H  G  S  F  D  D  S
```

Can You Create a New Sound Word Search of Your Own?

Write your Sound words in the grid below. Arrange them in the grid so they appear in horizontal (left to right), vertical (up or down), or diagonal (upward or downward) position. Fill in the blank boxes with other letters. Trade your Word Search with someone else who has created one of his or her own and see whether you can solve the new puzzle.

_____   _____   _____

_____   _____   _____

_____   _____   _____

_____   _____   _____

Hands-On Physical Science Activities

Answer Key for Sound Word Search

```
N B V C X V I B R A T E
E S M U S I C D F M G H
N Y U I O B P V L P K J
E H T M R R E O O L W Q
R E S O N A N C E I C Z
G A M L N T B A V F C X
Y R L E H I K L J Y H E
R E E C W O S C S D F G
T S T U Y N S O F T U I
C I O L V B N R U M P O
P O N E L O U D X N Z A
K N E S J H G S F D D S
```

Do You Recall?

Section Four: Sound

1. What does vibration have to do with sound?

2. What happens to pitch as the speed of vibration increases?

3. When you make sounds with your voice, what are some things that affect the "shape" of your voice? Which of these can you control as you speak or sing?

4. Consider a set of three soda bottles of the same size and shape, each containing a different amount of water:

   a. If you blow across the top of each one and make a tone, which bottle will produce the highest pitch, the one with the most water or the one with the least water?

Do You Recall? *(Cont'd.)*

b. If you tap each of the bottles with a spoon, which will produce the highest pitch, the one with the most water or the one with the least water?

5. What can you do to see vibrations that make sound?

6. Does sound travel faster through gases or through solids?

7. When there are fewer molecules, does sound travel better or not as well?

Answer Key for Do You Recall?

Section Four: Sound

Answer	Related Activities
1. All sounds come from vibrations.	All (especially 4.1)
2. The pitch goes up.	4.4, 4.6, 4.7
3. Answers will vary (shape of mouth, tongue, teeth, and so forth).	4.8, 4.9, 4.12
4. a. most water; b. least water	4.6
5. Answers will vary (for example, sprinkle rice on a drum and tap it)	4.1, 4.4, 4.10
6. Solids	4.13–4.16
7. Not as well	4.13–4.16

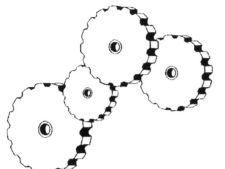

Simple Machines

To the Teacher

Acquiring an understanding of simple machines can help open our eyes to the world around us. All machines, regardless of complexity, are composed of various combinations of the six simple machines. These are often applied in unique and creative ways, but they are nonetheless the same six. After some exposure to these activities, students will enjoy applying their newly acquired awareness in identifying the simple machines in common appliances and equipment, such as the shovel, the eggbeater, the bicycle, the automobile, and so forth.

This section lends itself especially well to the discovery of scientific principles. Most of the activities suggested are safe for students to perform independently.

For most of the lever activities, a 1-inch board, which is approximately 1 m (1 yd.) long and 10 cm (4 in.) wide, is adequate. Others call for a lighter material, such as half-inch plywood.

It is recommended that you prepare your levers by marking positions 1, 2, 3, 4, and 5, measured at equal intervals, as indicated in Figure 5.A.

Figure 5.A. Lever with Points Marked and Eye Hooks

Eye hooks mounted at each point provide for attaching the spring balance.

Fulcrums ranging in height from 5 cm (2 in.) to 10 cm (4 in.) should be adequate and can be made by cutting a wedge shape from 4-inch by 4-inch post material, as shown in Figure 5.B. Scraps that are adequate can usually be acquired at a lumber store for little or no cost.

The following activities are designed as discovery activities that students can usually perform quite independently. You are encouraged to provide students (usually in small groups) with the materials listed and a copy of the activity from the beginning through the "Procedure." The section titled "Teacher Information" is not intended for student use, but rather to assist you with discussion following the hands-on activity, as students

Figure 5.B. 4- by 4-inch Fulcrum

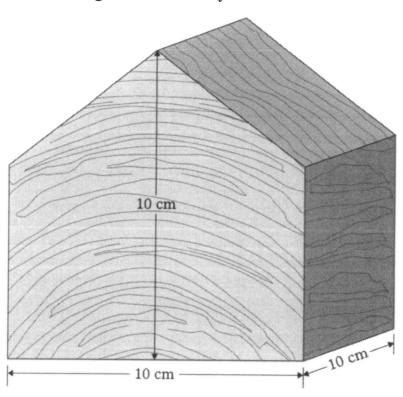

share their observations. Discussion of conceptual information prior to completing the hands-on activity can interfere with the discovery process.

Regarding the Early Grades

The first several activities can be easily adapted for younger children. By omitting the steps involving charting and mathematics, many of the later activities can be used as well. Young children can experience the concept "Machines make work easier" by feeling it and seeing it happen. These children can be encouraged to report in terms of "easier" and "harder" instead of by mathematical comparisons. For children who have experienced the teeter-totter (not commonly used today) and the wheelbarrow, these activities will help to clarify their earlier observations.

Correlation with National Standards

The following elements of the National Standards are reflected in the activities of this section.

K-4   Content Standard A: Science as Inquiry

As a result of activities in grades K-4, all students should develop:

1. Abilities necessary to do scientific inquiry
2. Understanding about scientific inquiry

K-4   Content Standard B: Physical Science

As a result of activities in grades K-4, all students should develop understanding of

1. Properties of objects and materials
2. Position and motion of objects
3. Light, heat, electricity, and magnetism

5-8   Content Standard A: Science as Inquiry

As a result of activities in grades 5-8, all students should develop:

1. Abilities necessary to do scientific inquiry
2. Understanding about scientific inquiry

5-8   Content Standard B: Physical Science

As a result of activities in grades 5-8, all students should develop understanding of

1. Properties and changes of properties in matter
2. Motions and forces
3. Transfer of energy

What Happens When You Rub Your Hands Together?

(Take home and do with family and friends.)

Materials Needed

- Copies of the "Science Investigation Journaling Notes" for this activity for all students

Procedure

1. As you complete this activity, you will keep a record of what you do, just as scientists do. Obtain a copy of the form "Science Investigation Journaling Notes" from your teacher and write the information that is called for, including your name and the date.

2. For this activity, you will learn what happens when you rub your hands together. For item 1, the question is provided for you on the form.

3. Item 2 asks for what you already know about the question. If you have some ideas about what happens when two objects rub together, write your ideas.

4. For item 3, write a statement of what happens when you rub your hands together, according to what you already know, and that will be your hypothesis.

5. Now continue with the following instructions. Complete your Journaling Notes as you go. Steps 6 through 9 below will help you with the information you need to write on the form for items 4 and 5.

6. Rub your hands briskly together for several seconds.

7. How do your hands feel?

8. Do it again, only faster. Then quickly hold your hands on your cheeks.

9. How do your hands feel to your cheeks?

10. Complete your "Science Investigation Journaling Notes." Are you ready to explain what happens when you rub your hands together? Discuss it with others.

Teacher Information

Whenever the surfaces of two objects rub together—the hands in this case—the resulting friction creates heat. In this simple activity, the heat is quickly noted and will vary according to the amount of moisture (perspiration and oil on the skin) that is present.

Integrating

Language arts

Science Process Skills

Observing, inferring, comparing and contrasting

Science Investigation
Journaling Notes for Activity 5.1

1. Question: *What happens when you rub your hands together?*

2. What we already know:

_____

_____

3. Hypothesis:

_____

_____

4. Materials needed:

_____

_____

5. Procedure:

_____

_____

6. Observations/New information:

_____

_____

7. Conclusion:

_____

_____

ACTIVITY 5.2

How Do Lubricants Affect Friction?

(Take home and do with family and friends.)

Materials Needed

- Pan of water or sink

Procedure

1. Rub your hands together briskly.
2. How do your hands feel?
3. Next, dip your hands in the water.
4. While they're wet, rub them briskly again.
5. Do your hands feel any different? Can you explain this?

For Problem Solvers

Do this project as a Science Investigation. Obtain a blank copy of the "Science Investigation Journaling Notes" from your teacher. Write your name, the date, and your question at the top. Plan your investigation through item 5 (Procedure) and have it approved by your teacher. Complete the Journaling Notes as you perform your investigation. Share your project with your group, and submit your Journaling Notes to your teacher if requested.

Put a small amount of olive oil, cooking oil, or hand lotion between your hands, and then rub your hands together again. Is it any easier to rub your hands together? Is there any difference in the amount of heat produced?

Find a fairly large standard screwdriver (flat tip). Hold it by the bit, the end that slants and fits the screw. Next, wet your fingers and hold it by the bit again. Is it just as easy to hold as before? Wipe the water off your fingers and the screwdriver; then put a drop of cooking oil between your fingers and hold the screwdriver the same way. What difference do you find? Explain why.

Teacher Information

In addition to providing a cooling effect, the water also acts as a lubricant, reducing friction and thereby reducing the amount of heat produced by friction.

Your problem solvers will find that friction is greatly reduced by the use of a lubricant, such as cooking oil, olive oil, or hand lotion. If motor oil is available, let them try it with the screwdriver activity and compare with the other lubricants.

Integrating

Language arts

Science Process Skills

Observing, inferring, communicating, comparing and contrasting, identifying and controlling variables, experimenting

Hands-On Physical Science Activities

How Do Starting Friction and Sliding Friction Compare?

(Take home and do with family and friends.)

Materials Needed
- Two or three large books
- String 2 m (2 yds.) long
- Spring balance

Procedure
1. Tie the books into a bundle, using the string.
2. Place the books on a table or on the floor. Attach one end of the spring balance to the string wrapped around the books, as shown in Figure 5.3–1.

Figure 5.3–1. Setup for Activity 5.3

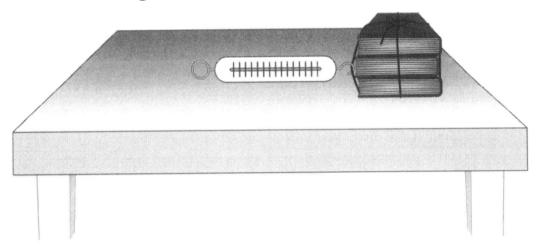

3. Holding the other end of the spring balance and watching the indicator needle carefully, pull the books 50 to 100 cm (1.5 to 3 ft.) across the table (or floor).

4. What was the reading on the spring balance when the books first began to move?

5. What was the reading on the spring balance as the books moved steadily across the table?

6. Repeat the activity, being sure to pull the books in a steady, not jerking, manner.

7. Is the amount of force required to start the books moving equal to the amount of force needed to keep them moving? If not, which is greater?

8. Repeat to verify your findings.

9. Discuss your findings with others.

For Problem Solvers

Do this project as a Science Investigation. Obtain a blank copy of the "Science Investigation Journaling Notes" from your teacher. Write your name, the date, and your question at the top. Plan your investigation through item 5 (Procedure) and have it approved by your teacher. Complete the Journaling Notes as you perform your investigation. Share your project with your group, and submit your Journaling Notes to your teacher if requested.

Do you think friction will be the same for a stack of books if a small book is at the bottom of the pile as it will be if a large book is at the bottom? If you don't think it will be the same, which do you think will have the least friction? Do some scientific investigation and find out. Think through your procedure and discuss it with your teacher or someone else before you begin. Test the question for both starting friction and sliding friction.

Continue your investigation to compare the friction of various surfaces rubbing against each other. What could you put under the stack of books that you think would allow the books to slide across the table with less friction? Try at least two or three different materials and remember to test both starting friction and sliding friction. Make a graph of your results.

Teacher Information

Starting friction is greater than sliding friction. More force is required to start an object than to keep it sliding. One factor is inertia, the tendency of an object at rest to remain at rest and of an object in motion to remain in motion.

Those who accept the "For Problem Solvers" challenge will practice their skills at designing a simple experiment, interpreting the data, and communicating their results in the form of a graph. They will also learn some useful information about friction and the difference that is made by the type of surfaces involved.

Integrating

Math

Science Process Skills

Observing, inferring, measuring, predicting, communicating, comparing and contrasting, formulating hypotheses, identifying and controlling variables, experimenting

How Does Rolling Friction Compare with Sliding Friction?

(Take home and do with family and friends.)

Materials Needed

- Two or three large books
- String 2 m (2 yd.) long
- Spring balance
- At least six round pencils
- Copies of the "Science Investigation Journaling Notes" for this activity for all students

Procedure

1. As you complete this activity, you will keep a record of what you do, just as scientists do. Obtain a copy of the form "Science Investigation Journaling Notes" from your teacher and write the information that is called for, including your name and the date.

2. For this activity, you will learn how rolling friction compares with sliding friction. For item 1, the question is provided for you on the form.

3. Item 2 asks for what you already know about the question. If you have some ideas about friction and why we use wheels, write your ideas.

4. For item 3, write a statement of how sliding and rolling friction compare, according to what you already know, and that will be your hypothesis.

5. Now continue with the following instructions. Complete your Journaling Notes as you go. Steps 6 through 14 below will help you with the information you need to write on the form for items 4 and 5.

6. Tie the books into a bundle, using the string.

7. Place the bundle of books on a table or on the floor.

8. Attach one end of the spring balance to the string wrapped around the books.

9. While holding the other end of the spring balance and watching the indicator needle carefully, slide the books steadily 25 to 50 cm (10 to 20 in.) across the floor (or table).

Hands-On Physical Science Activities

10. Record the amount of force needed for both starting friction and sliding friction.

11. Next, place the pencils side by side, about 5 to 8 cm (2 to 3 in.) apart.

12. Place the books on the pencils at one end of the row.

13. Pull the books again with the spring balance and record the amount of force required for starting friction and rolling friction.

14. Did the pencils change the force needed to drag the books across the table? If so, how much difference did they make?

15. Complete your "Science Investigation Journaling Notes." Are you ready to explain how sliding and rolling friction compare? Discuss it with your group.

For Problem Solvers

Do the following project also as a Science Investigation. Obtain another blank copy of the "Science Investigation Journaling Notes." Write your name, the date, and your question at the top. Plan your investigation through item 5 (Procedure) and have it approved by your teacher. Complete the Journaling Notes as you perform your investigation. Share your project with your group, and submit your Journaling Notes to your teacher if requested.

Predict whether the starting friction and rolling friction will be the same with the books on a skateboard as with the pencils. Place the same books on a skateboard and compare both starting friction and rolling friction with the results you got when using the pencils. Add your skateboard data to your graph.

Teacher Information

Rolling friction is less than sliding friction. This principle is used in wheels and bearings in a wide variety of applications, from wheels under the legs of a table to the workings of complex machinery. The Egyptians probably used rollers to move large stones when they built the pyramids.

Integrating

Math

Science Process Skills

Observing, inferring, measuring, predicting, communicating, comparing and contrasting, formulating hypotheses, identifying and controlling variables, experimenting

Science Investigation
Journaling Notes for Activity 5.4

1. Question: *How does rolling friction compare with sliding friction?*

2. What we already know:

_____

_____

3. Hypothesis:

_____

_____

4. Materials needed:

_____

_____

5. Procedure:

_____

_____

6. Observations/New information:

_____

_____

7. Conclusion:

_____

_____

What Is the Advantage of a First-Class Lever?

(Take home and do with family and friends.)

Materials Needed

- Board 10 cm (4 in.) wide and 1 m (1 yd.) long (or other suitable lever)
- Fulcrum
- Book
- Copies of the "Science Investigation Journaling Notes" for this activity for all students

Procedure

1. As you complete this activity, you will keep a record of what you do, just as scientists do. Obtain a copy of the form "Science Investigation Journaling Notes" for this activity from your teacher and write the information that is called for, including your name and the date.

2. For this activity, you will learn about how we benefit by using first-class levers. For item 1, the question is provided for you on the form.

3. Item 2 asks for what you already know about the question. If you know something about levers, and particularly the first-class lever, write your ideas.

4. For item 3, write a statement of the advantages of the first-class lever, according to what you already know, and that will be your hypothesis.

5. Now continue with the following instructions. Complete your Journaling Notes as you go. Steps 6 through 14 below will help you with the information you need to write on the form for items 4 and 5.

6. Place the fulcrum under the lever (board) at the middle (position 3 in Figure 5.5–1).

Figure 5.5–1. Lever on Fulcrum, Arrow Marking the Effort Point

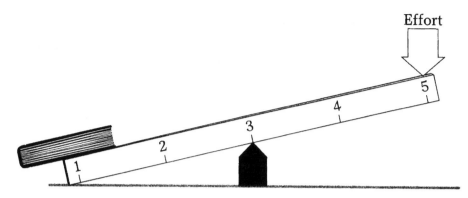

7. Place the book on the lever at position 1. Push down on the lever at position 5.

8. What happened to the book as you pushed down on the other end of the lever?

9. Repeat this procedure with the fulcrum at position 2 and again at position 4.

10. Is the same amount of effort required to raise the book regardless of the fulcrum position?

11. How does the required effort change as you move the fulcrum away from the book?

12. How does the required effort change as you move the fulcrum closer to the book?

13. Try to lift some other objects with your lever, such as a box of copy paper. *Warning:* Do not try to lift a piano or other objects that are heavy and tall, as they might tip over.

14. To lift a heavy object, would you place the fulcrum near the object, away from the object, or does it matter?

15. Complete your "Science Investigation Journaling Notes." Are you ready to explain the advantages of a first-class lever? Discuss it with your group.

For Problem Solvers

Look around you at school, at home, and everywhere else you go for the next several days. Find every example of levers that you can. Make a list of these. You will use your list as you do some of the next activities. Continue to add to your list as you go.

Teacher Information

The fulcrum divides a first-class lever into two parts, called the effort arm and the load arm. The load arm is the end upon which the load rests. The effort arm is the end we apply a force to in order to move the load.

With the fulcrum between the load arm and the effort arm, the first-class lever changes the direction of the load; that is, we move the load up by pressing down on the opposite end of the lever.

With a first-class lever, moving the fulcrum toward the load decreases the amount of force necessary to lift the load. When the fulcrum is closer to the load, we gain in force but lose speed and distance. When the fulcrum is closer to the effort, we gain in speed and distance but lose force.

Examples of first-class levers include scissors, pliers, and teeter-totters.

Warning: Do not use a piano as an object to lift with a lever.

Integrating

Math

Science Process Skills

Observing, inferring, measuring, predicting, communicating, formulating hypotheses, identifying and controlling variables, experimenting

Science Investigation

Journaling Notes for Activity 5.5

1. Question: *What is the advantage of a first-class lever?*

2. What we already know:

_____

_____

3. Hypothesis:

_____

_____

4. Materials needed:

_____

_____

5. Procedure:

_____

_____

6. Observations/New information:

_____

_____

7. Conclusion:

_____

_____

What Type of Simple Machine Is the Teeter-Totter?

(Take home and do with family and friends.)

Materials Needed

- Teeter-totter (improvised from a plank and a fulcrum)

Procedure

1. Place the teeter-totter in such a position that it will balance with you on one end and a friend on the other end.
2. Change the position of the teeter-totter on the fulcrum (bar in the middle) and try to balance with the same person.
3. What happened?
4. Now adjust the teeter-totter so you can balance with a different friend.
5. What did you have to do? Why?

For Problem Solvers

You might not have thought of the teeter-totter as a machine until now, because teeter-totters are used for entertainment. Well, many toys and other things we use for entertainment are machines. The teeter-totter is a first-class lever. As a machine, it allows a small person to lift a large person.

Place the teeter-totter in the middle, so it balances (or almost) by itself. Have two people get on it who are about the same size. Did it balance?

Find two people who are different sizes and who are willing to help you. They can be different ages. Predict where the teeter-totter needs to be placed on the bar in order for it to balance with these two people. Try it, and check your prediction. Do the same thing with several more pairs of people and try to improve your accuracy with each prediction.

Teacher Information

Although teeter-totters are no longer a common piece of playground equipment, one can easily be prepared using a plank and a fulcrum.

The teeter-totter is a first-class lever. Either end could be called the effort arm or the load arm. This is an excellent example for reinforcing the concept of center of gravity.

Integrating

Math

Science Process Skills

Observing, inferring, measuring, predicting, communicating, comparing and contrasting, formulating hypotheses, identifying and controlling variables, experimenting

How Can a Lever Be Used to Lift Heavy Things?

(Teacher demonstration)

Materials Needed

- Lever (2-in. plank)
- Fulcrum
- Automobile

Procedure

1. Place the plank on the ground and drive an automobile into such a position that one tire is on one end of the plank.
2. Set the parking brake.
3. Lift the end of the plank opposite the wheel and place the fulcrum under the plank near the wheel.
4. Push down on the effort arm of the lever.
5. Ask students what you can do with the lever that you couldn't do without it.

Teacher Information

Caution: This activity is suggested as a teacher demonstration because of the obvious potential risks involved. It demonstrates that levers can be used to lift very heavy loads with relatively little effort. Perhaps student assistants could be used safely for some tasks, but care should be taken to avoid risk of injury.

Consider having students try to lift one side of the automobile before using the lever. Be very careful because of possible back injuries. One student can lift a corner of the car using a lever, while several students could not do it without the lever.

Integrating

Math

Science Process Skills

Observing, inferring, measuring, predicting, communicating, formulating hypotheses, identifying and controlling variables, experimenting

How Can You Predict the Effort Required to Lift a Load with a First-Class Lever?

Materials Needed

- Lever 1 m (1 yd.) long (preferably lightweight, such as half-inch plywood)
- String
- Fulcrum 3 to 10 cm high (1 to 4 in.)
- Spring balance
- Two or three books
- "Record of Measurement Chart I" for each student
- Pencils

Procedure

1. Tie the books into a bundle. Place the fulcrum under position 3 as indicated in Figure 5.8–1.
2. Place the books on the lever at position 1.
3. Use the "Record of Measurement Chart I" for recording your measurements in this activity.
4. Attach the spring balance at position 5. Pull down and record the force required to lift the books.
5. Weigh the books and compare with the force required to lift them in step 4.

Figure 5.8–1. Lever with Fulcrum at Position 3

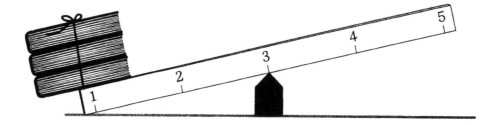

6. With the books on the lever at position 1 and the spring balance attached at position 5, move the fulcrum to position 4.

7. Pull down on the spring balance and record the force required to lift the books.

8. Repeat the above procedure with the fulcrum at position 2.

9. Compare your findings.

10. Predict the force required to lift the books with the fulcrum halfway between positions 2 and 3. Record your prediction.

11. Try it. Record the actual force required. How close was your prediction?

12. Predict the force necessary to lift the books with the fulcrum halfway between positions 3 and 4. Record your prediction.

13. Try it. Record your results. Was your prediction any closer this time?

Teacher Information

The effort required to lift an object with the first-class lever is proportionate to the relative lengths of the load arm and the effort arm. For example, with the fulcrum at position 3, the two arms are equal in length. If the load weighs 1 kg (or 1 lb.), the effort required to lift it should be 1 kg (or 1 lb.). (*Note:* 1 kg = 2.2 pounds, but load and effort are equal with the fulcrum at position 3.)

With the fulcrum in position 4, the load arm is three times as long as the effort arm and the effort required to lift 1 kg (or 1 lb.) will be about 3 kg (or 3 lbs.).

With the fulcrum in position 2, the effort arm is three times as long as the load arm and the effort required to lift 1 kg (or 1 lb.) will be about 0.33 kg (or .33 lb.).

The degree of accuracy of the figures is affected by the degree of precision in positioning the load, fulcrum, and effort, and by the weight of the board itself. The results are therefore only approximate.

Integrating

Math

Science Process Skills

Observing, inferring, measuring, predicting, communicating, comparing and contrasting, formulating hypotheses, identifying and controlling variables, experimenting

Name _____ Date _____

Record of Measurement Chart I

Actual weight of the load = _____ kg

Load Position	Effort Position	Fulcrum Position	Force
1	5	3	_____
1	5	4	_____
1	5	2	_____
1	5	between 2 and 3	Prediction: _____ Actual: _____
1	5	between 3 and 4	Prediction: _____ Actual: _____

What Do We Lose as We Gain Force with a Lever?

Materials Needed

- Lever 1 m (1 yd.) long (preferably lightweight, such as half-inch plywood)
- Pencils
- Spring balance
- Fulcrum at least 10 cm (4 in.) high
- Two or three books
- String
- "Record of Measurement Chart II" for each student
- Rulers

Procedure

1. Use the "Record of Measurement Chart II" for recording your measurements in this activity.
2. Tie the books into a bundle. Place the fulcrum under position 3, as indicated in Figure 5.9–1.

Figure 5.9–1. Lever with Fulcrum at Position 3

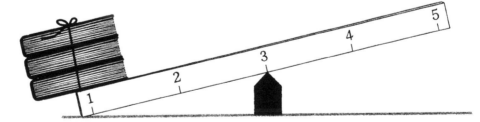

Simple Machines

3. In addition to measuring the force required to lift the load for the various fulcrum positions, use the ruler to measure the distances traveled by the effort arm while the load arm travels the distances indicated on the chart.

    a. Place the books at position 1.

    b. Attach the spring balance at position 5, pull down, and record the force required to lift the books.

    c. Weigh the books and compare with the force required above.

    d. With the books on the lever at position 1 and the spring balance attached at position 5, move the fulcrum to position 4. (See Figure 5.9–2.)

    e. Pull down on the spring balance and record the force required to lift the books.

    f. Repeat the above procedure with the fulcrum at position 2.

    g. Compare your findings.

4. With the last two fulcrum positions, predict travel distances of the effort arm. Record your predictions.

5. Try these and record the actual results.

6. As the force required at the effort arm decreases, does the distance the effort arm travel increase or decrease?

7. Write a statement about the force required to lift a load, the distance the load travels, and the distance the effort arm travels as the fulcrum is moved closer and closer to the load.

8. The lever you used here is called a first-class lever. Notice how it compares to the second-class lever and the third-class lever as you do other activities.

Figure 5.9–2. Lever with Fulcrum at Position 4

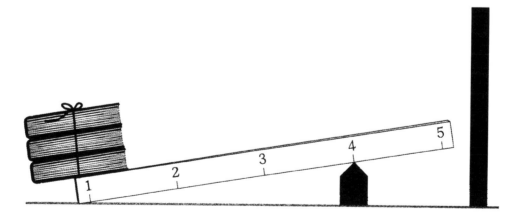

Hands-On Physical Science Activities

For Problem Solvers

If you did the "For Problem Solvers" in Activity 5.5, examine the list of levers you made at that time. If not, look around for the next few days and make a list of all of the examples of levers you can find at home, at school, and wherever you go. From your list, identify those that are first-class levers. For each one, decide whether the advantage of using the lever in this application is to gain force or to gain distance and speed.

Continue to watch for more applications of levers of all kinds. Add them to your list as you find them.

Teacher Information

The total amount of work required to lift a load is neither increased nor decreased by the use of a lever. In using a first-class lever (as in this activity), we can decrease the amount of force required to lift a load by moving the fulcrum closer to the load. As the fulcrum moves closer to the load and the effort required to lift the load is decreased, the effort arm travels a greater distance and the load travels a lesser distance. We gain in terms of force required, but we sacrifice speed and distance.

The amount of force required to lift a given load using a first-class lever can be computed using the following formula:

load (length of load arm) = effort (length of effort arm)

For example, if we have a 200-kg load and we can apply only 50 kg of force to lift the load, the effort arm must be four times as long as the load arm. As the effort required to lift the load is divided by four, the speed and distance traveled by the load will also be divided by four.

In their list of first-class levers, your problem solvers should note that the advantage can go either way, depending on the position of the fulcrum. First-class levers can provide gain in force with a sacrifice of distance and speed, or a gain in distance and speed with a sacrifice of force. If the fulcrum is closer to the load, the gain is in force. If the fulcrum is closer to the effort position, the gain is in distance and speed. If the fulcrum is in the center of the lever, the only advantage is that it reverses the direction of the load. The first-class lever always reverses the direction of the load. For each example in their lists, students should be able to easily determine the advantage by noting the position of the fulcrum.

Integrating

Math

Science Process Skills

Observing, inferring, measuring, predicting, communicating, comparing and contrasting, formulating hypotheses, identifying and controlling variables, experimenting

Record of Measurement
Chart II

Weight of the load = _____ kg

Load Position	Effort Position	Fulcrum Position	Force	Travel Distance Load Arm	Travel Distance Effort Arm
1	5	3	_____	10 cm	_____
1	5	4	_____	5 cm	_____
1	5	2	_____	5 cm	_____
1	5	Between 2 and 3	Prediction: _____ Actual: _____	5 cm	Prediction: _____ Actual: _____
1	5	Between 3 and 4	Prediction: _____ Actual: _____	10 cm	Prediction: _____ Actual: _____

Simple Machines

How Is a Second-Class Lever Different from a First-Class Lever?

Materials Needed

- Lever
- Fulcrum
- Two or three books
- String
- Copies of the "Science Investigation Journaling Notes" for this activity for all students

Procedure

1. As you complete this activity, you will keep a record of what you do, just as scientists do. Obtain a copy of the form "Science Investigation Journaling Notes" from your teacher and write the information that is called for, including your name and the date.

2. For this activity, you will learn about second-class levers, and how they are different from first-class levers. For item 1, the question is provided for you on the form.

3. Item 2 asks for what you already know about the question. If you know something about different types of levers, write your ideas.

4. For item 3, write a statement of how second-class levers are different from first-class levers, according to what you already know, and that will be your hypothesis.

5. Now continue with the following instructions. Complete your Journaling Notes as you go. Steps 6 through 12 below will help you with the information you need to write on the form for items 4 and 5.

6. Tie the books into a bundle.

7. Place the fulcrum at position 5 and hang the books, by their string, from position 4, as illustrated in Figure 5.10–1.

Figure 5.10–1. Lever, Books, and Fulcrum

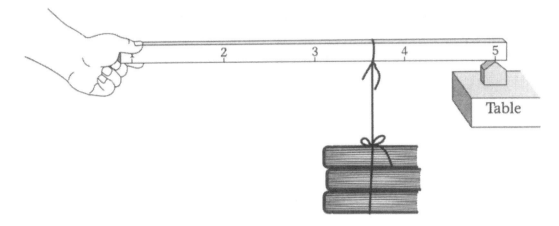

8. Holding the lever at position 1, lift the books.

9. Move the books to position 3, then 2, then 1, each time lifting from position 1.

10. Is it easier to lift when the load is closer to the fulcrum or farther from the fulcrum?

11. With the books at position 4, the fulcrum still at position 5, and the effort still applied at position 1, what is the length of the load arm? The effort arm?

12. This is a second-class lever. Notice the relative positions of the fulcrum, the load, and the effort for this second-class lever. How do these compare with the first-class lever you used?

13. Complete your "Science Investigation Journaling Notes." Are you ready to explain how second-class levers are different from first-class levers? Discuss it with your group.

Teacher Information

With the load placed between the fulcrum and the effort, we now have a second-class lever. The length of each arm is measured from the fulcrum, so with the load at position 3, the effort arm is twice the length of the load arm.

As with the first-class lever, the shorter the load arm and the longer the effort arm, the less effort required to lift the load. The effort arm travels farther and faster, however, than the load.

A major difference between the first-class lever and the second-class lever is that the second-class lever does not reverse the direction of the load; both effort and load travel in the same direction.

Examples of second-class levers include the paper cutter, the nutcracker, and the wheelbarrow.

Integrating

Math

Science Process Skills

Observing, inferring, measuring, predicting, communicating, comparing and contrasting, formulating hypotheses, identifying and controlling variables, experimenting

Hands–On Physical Science Activities

Science Investigation

Journaling Notes for Activity 5.10

1. Question: *How is a second-class lever different from a first-class lever?*

2. What we already know:

_____

_____

3. Hypothesis:

_____

_____

4. Materials needed:

_____

_____

5. Procedure:

_____

_____

6. Observations/New information:

_____

_____

7. Conclusion:

_____

_____

Simple Machines

What Do You Gain and What Do You Lose by Using a Second-Class Lever?

Materials Needed

- Lever
- Fulcrum
- Two or three books
- String
- Pencil
- Spring balance
- Meter stick
- "Record of Measurement Chart III" for each student

Procedure

1. Use the "Record of Measurement Chart III" for recording your measurements in this activity.
2. Tie the books into a bundle.
3. Weigh the books and record the results.
4. Place the fulcrum at position 5 and suspend the books, by the string, from position 1, as shown in Figure 5.11-1.
5. Record the amount of force required to lift the books, with the spring balance also at position 1, and compare this force to the weight of the books.
6. Measure the distance traveled by the spring balance (effort) as the books (load) travel 20 cm (8 in.).
7. Next, move the load to position 3 and record the force indicated on the spring balance. With the load between effort and fulcrum, you now have a second-class lever.
8. How does the amount of force required compare with the actual weight of the books?

Figure 5.11–1. Lever with Books, Fulcrum, and Positions Noted

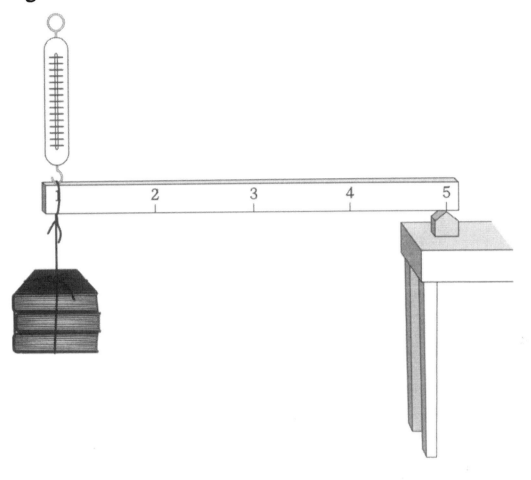

9. Lift the load, measuring the distance traveled by the effort as the load is raised 10 cm (4 in.). Record the results.

10. How does the distance traveled by the load arm compare with the distance traveled by the effort arm?

11. If you were to move the load to position 4, how much force do you think would be required to lift the books? Record your prediction.

12. Try it and record the results. How close was your prediction?

13. With the load at position 2, how far do you think the load will travel as you lift the effort 20 cm (8 in.)? Record your prediction.

14. Try it and record the results. How close was your prediction?

For Problem Solvers

Examine your list of levers and find all of them that are second-class levers. For each one, decide whether the advantage of using the lever in this application is to gain force or to gain distance and speed.

Continue to watch for more applications of the lever. Add them to your list as you find them.

Teacher Information

The formula for computing effort and travel distance is the same for the second-class lever as for the first-class lever. Remember to measure the length of each arm from the fulcrum.

All second-class levers provide a gain in force with a sacrifice of distance and speed. In their list of second-class levers, your problem solvers should have included such things as the wheelbarrow, the nutcracker, and the paper cutter. The advantage of each one is in force.

Integrating

Math

Science Process Skills

Observing, inferring, measuring, predicting, communicating, comparing and contrasting, formulating hypotheses, identifying and controlling variables, experimenting

Record of Measurement Chart III

Weight of the load = _____ kg

Load Position	Effort Position	Fulcrum Position	Force	Travel Distance Load Arm	Effort Arm
				Load Arm	**Effort Arm**
1	1	5	_____	20 cm	_____
3	1	5	_____	10 cm	_____
4	1	5	Prediction: _____ Actual: _____	Prediction: _____ Actual: _____	20 cm
2	1	5	Prediction: _____ Actual: _____	Prediction: _____ Actual: _____	20 cm

Simple Machines

What Is a Third-Class Lever?

Materials Needed

- Lever
- Table
- Two or three books
- Strings
- Copies of the "Science Investigation Journaling Notes" for this activity for all students

Procedure

1. As you complete this activity, you will keep a record of what you do, just as scientists do. Obtain a copy of the form "Science Investigation Journaling Notes" from your teacher and write the information that is called for, including your name and the date.

2. For this activity, you will learn what a third-class lever is and how it is different from first- and second-class levers. For item 1, the question is provided for you on the form.

3. Item 2 asks for what you already know about the question. If you know something about third-class levers and why we use them, write your ideas.

4. For item 3, write a statement of what a third-class lever is, according to what you already know, and that will be your hypothesis.

5. Now continue with the following instructions. Complete your Journaling Notes as you go. Steps 6 through 15 below will help you with the information you need to write on the form for items 4 and 5.

6. Tie the books into a bundle and weigh them.

7. Use the edge of the table as a fulcrum. (You might need to have someone sit on the table to hold it down.)

8. Place the end of the lever under the edge of the table so your fulcrum (table's edge) is at position 5.

9. Suspend the books, by their string, at position 1, as shown in Figure 5.12–1.

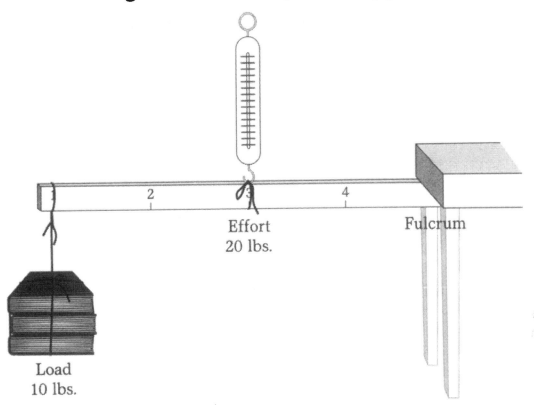

Figure 5.12–1. Setup for Activity 5.12

Effort
20 lbs.

Fulcrum

Load
10 lbs.

10. Holding the lever at position 3, lift the books.

11. Is the effort required to lift the books greater or less than the actual weight of the books?

12. Move your hand to position 2 and lift the load.

13. Lift the load from position 4.

14. Is it easier to lift the load as the effort (your hand) moves closer to the fulcrum (the table's edge)?

15. How is the third-class lever different from the first-class and second-class levers you have been using?

16. Complete your "Science Investigation Journaling Notes." Are you ready to explain what a third-class lever is? Discuss it with your group.

Teacher Information

As explained in earlier activities, second-class levers decrease the amount of effort required to lift a load, but in so doing they increase the distance the effort must travel to lift the load a given distance. The third-class lever reverses the advantage. With the effort now between the fulcrum and the load, the effort required to lift the load is greater than the actual weight of the load. The load, however, travels faster and farther than does the effort.

As with the other types of levers, the lengths of both the effort arm and the load arm are measured from the fulcrum. The load arm is always longer than the effort arm with a third-class lever.

The speed-and-distance advantage of the third-class lever is helpful in the use of such items as the fishing pole, ax, and broom. Our arms and legs are also third-class levers, with the joint as the fulcrum. The distance from joint to hand or foot is the load arm. The effort arm is from the joint to the point at which the tendons attach to anchor the muscle to the bone. These third-class levers offer advantages in speed and distance as a person swings a bat or a golf club, throws a baseball, or kicks a soccer ball.

Integrating

Math

Science Process Skills

Observing, inferring, measuring, predicting, communicating, comparing and contrasting, formulating hypotheses, identifying and controlling variables, experimenting

Science Investigation

Journaling Notes for Activity 5.12

1. Question: *What is a third-class lever?*

2. What we already know:

_____

_____

3. Hypothesis:

_____

_____

4. Materials needed:

_____

_____

5. Procedure:

_____

_____

6. Observations/New information:

_____

_____

7. Conclusion:

_____

_____

What Is Gained and What Is Lost by Using a Third-Class Lever?

Materials Needed

- Lever
- Table
- Two or three books
- String
- Pencil
- Spring balance
- Meter stick
- "Record of Measurement Chart IV" for each student

Procedure

1. Use the "Record of Measurement Chart IV" for recording your measurements in this activity.
2. Tie the books into a bundle.
3. Weigh the books and record the weight.
4. Use the edge of the table as a fulcrum. (You might need to have someone sit on the table to hold it down.)
5. Place the end of the lever under the edge of the table so your fulcrum (table's edge) is at position 5. (Refer to Figure 5.12–1 in Activity 5.12.)
6. Suspend the books, by their string, at position 1.
7. Attach one end of the spring balance to the lever at position 3.
8. Holding the other end of the spring balance, lift with enough force to support the books. You are using a third-class lever.
9. Record the reading at the indicator and compare with the actual weight of the books.
10. Lift the load from position 3 and measure the travel distance of the load as the effort travels 10 cm.

322

11. Using the prediction skills you have learned in earlier activities, predict the force required to lift the load with the effort being shifted to position 2.

12. Try it, record the results, and compare with your prediction.

13. Leaving the effort at position 2, predict the travel distance of the load as the effort travels 10 cm.

14. Next, predict the outcomes with the effort being applied at position 4 and the effort traveling 5 cm, and record the results.

15. Try it. Were your predictions close?

16. Select another point on the lever, somewhere between the numbers. Predict effort and distances and test your predictions.

For Problem Solvers

Examine your list of levers and find all of them that are third-class levers. For each one, decide whether the advantage of using the lever in this application is to gain force or to gain distance and speed.

Teacher Information

The formula for computing effort and travel distances for a third-class lever is the same as for first- and second-class levers. Remember to measure the lengths of the effort and load arms from the fulcrum.

With the system set up as indicated above, the effort required to lift a 10-pound load would be 20 pounds, since the load arm is twice the length of the effort arm.

All third-class levers provide gain in distance and speed with a sacrifice of force. In their list of third-class levers, your problem solvers should have included such things as the fishing pole, the ball bat, ax, broom, golf club, and their own arms and legs. The advantage of each one is in distance and speed.

Integrating

Math

Science Process Skills

Observing, inferring, measuring, predicting, communicating, comparing and contrasting, formulating hypotheses, identifying and controlling variables, experimenting

Record of Measurement Chart IV

Weight of the load = _____ kg

Load Position	Effort Position	Fulcrum Position	Force	Travel Distance	
				Load Arm	**Effort Arm**
1	3	5	_____	_____	10 cm
1	2	5	Prediction: _____ Actual: _____	Prediction: _____ Actual: _____	10 cm
1	4	5	Prediction: _____ Actual: _____	Prediction: _____ Actual: _____	5 cm
1	?	1	Prediction: _____ Actual: _____	Prediction: _____ Actual: _____	5 cm

Copyright © 2006 by John Wiley & Sons, Inc.

What Is the Wheel and Axle?

Materials Needed

- Compasses
- Stiff paper at least 10 cm (4 in.) square
- Pencils
- Scissors
- Tape measures

Procedure

1. Use a compass to make a circle on a piece of the paper.
2. Cut out the circle.
3. Insert a pencil through the center of the circle. You have made a wheel and axle, just like Figure 5.14–1.
4. Roll your wheel and axle along the tabletop. How many times does the pencil rotate as the wheel rotates once?
5. Measure the distance the wheel traveled in one complete rotation.

Figure 5.14–1. Pencil and Paper Wheel on Table

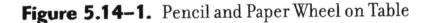
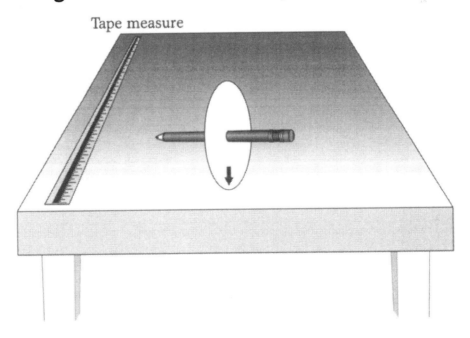

Simple Machines

6. Remove the pencil from the wheel, place the pencil on the table, and measure the distance it travels in one complete rotation.

7. How far would the pencil travel if rotated ten times?

8. Insert the pencil (axle) back into the wheel. How far does it travel now in ten rotations?

9. Name one advantage of the wheel and axle.

Teacher Information

The wheel and axle is a form of lever. When the wheel or the axle turns, the other turns also. If the wheel turns around the axle, as on bearings, it is not a "wheel and axle" in terms of the concepts of simple machines.

If the wheel is turning the axle, it is a form of second-class lever. (See Figure 5.14–2.) The fulcrum is at the center of the axle. The radius of the wheel is the effort arm of the lever, and the radius of the axle is the load arm. There is increased force but less speed and distance.

Examples of the wheel and axle acting as a second-class lever include the doorknob, the screwdriver, and the steering wheel of an automobile.

Figure 5.14–2. Wheel and Axle Showing Load and Effort as Second-Class Lever

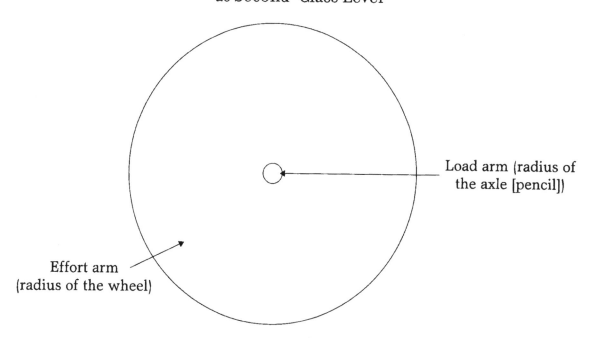

Load arm (radius of the axle [pencil])

Effort arm (radius of the wheel)

Figure 5.14–3. Wheel and Axle Showing Load and Effort
as Third-Class Lever

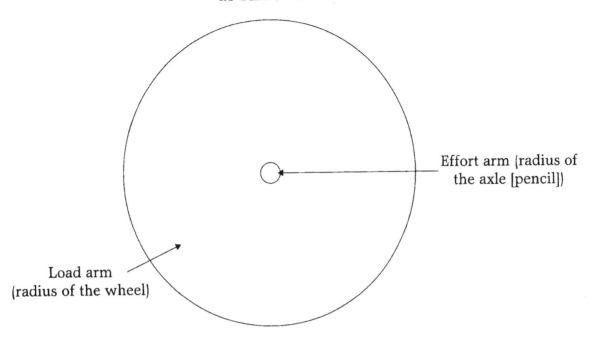

Effort arm (radius of
the axle [pencil])

Load arm
(radius of the wheel)

If the axle is turning the wheel, it becomes a form of third-class lever, with a gain in speed and distance but a decrease in force. (See Figure 5.14–3.) The fulcrum is at the center of the axle. The radius of the wheel is the load arm and the radius of the axle is the effort arm.

Examples of the wheel and axle acting as a third-class lever include the drive wheels of an automobile and the rear wheel (drive wheel) of a bicycle.

Integrating

Math

Science Process Skills

Observing, inferring, measuring, predicting, communicating, comparing and contrasting, formulating hypotheses, identifying and controlling variables, experimenting

Simple Machines

What Type of Simple Machine Is the Pencil Sharpener?

Materials Needed

- Empty pencil sharpeners with suction mounts
- Pieces of string about 1 m (1 yd.) long
- Small books (or other light weight)
- Masking tape

Procedure

1. Clamp the pencil sharpener to the side of a file cabinet or other vertical surface and remove the cover.

2. Turn the handle of the pencil sharpener around, noting that it goes all the way around, just like a wheel.

3. Tie a small book (or other light weight) in such a way that a long string is left from which the book can be suspended.

4. Notice the amount of effort required to lift the book. Tie the end of the string firmly around the end of the pencil sharpener shaft. Use tape to keep it from slipping. The result should resemble Figure 5.15–1.

5. Allow the book to hang freely and support the pencil sharpener with your hands to keep it from pulling loose.

6. Turn the pencil sharpener handle around several times, making sure the string is winding around the shaft.

7. Is more or less force required to lift the book with this wheel and axle than to lift the book directly?

8. See whether you can locate a picture of an old well with a windlass for raising a bucket full of water. What similarities do you see between the windlass and your pencil sharpener?

9. What type of machine is the pencil sharpener? The windlass?

Teacher Information

A wheel does not have to be a complete wheel in order to be considered a wheel and axle. It can be just a crank, as with the pencil sharpener used above or the water-well windlass referred to. The crank makes a complete

328

Figure 5.15–1. Setup for Activity 5.15

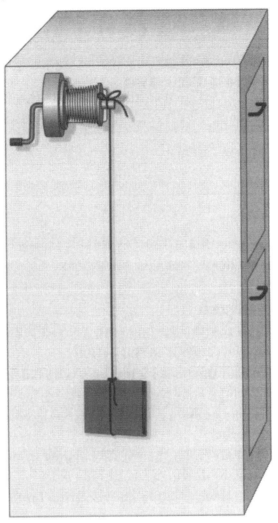

circle when used, just as though it were a complete wheel. A type of windlass called a winch is often found on boat trailers and four-wheel-drive vehicles.

Integrating

Math

Science Process Skills

Observing, inferring, measuring, predicting, communicating, comparing and contrasting, formulating hypotheses, identifying and controlling variables, experimenting

Simple Machines

What Is a Fixed Pulley?

Materials Needed

- Single-wheel pulley
- Crossbar
- Spring balance
- Meter stick (or yardstick)
- Cord or heavy string
- Pencils
- "Measuring with a Fixed Pulley" chart for each student
- Bundle of books (or other heavy object)

Procedure

1. Use the "Measuring with a Fixed Pulley" chart for recording your measurements in this activity.

2. Weigh the books with the spring balance and record the results.

3. Arrange your pulley, crossbar, spring balance, cord, and bundle of books as shown in Figure 5.16–1, with the pulley attached to the crossbar.

4. Pull down on the spring balance to lift the books. Be sure to pull straight down and not to the side.

5. Pull down steadily on the spring far enough to lift the load 20 cm. Record the following information on your chart:

   a. Direction the load (books) moved as the effort (spring balance) moved downward.

   b. Distance moved by the effort as the load moved 20 cm.

   c. Amount of force required to lift the books.

6. Examine the information in your chart. Was lifting the books using the pulley different in any way from lifting the books without the pulley? Consider these questions:

   a. Did the pulley decrease the amount of force needed to lift the books?

   b. Did the pulley cause the books to move a greater or lesser distance than the effort moved?

   c. Did the pulley cause the books to move in the opposite direction from that of the effort?

Hands-On Physical Science Activities

7. With the pulley fastened to the crossbar, as it has been for this activity, it is called a fixed pulley. This simply means that the pulley does not move up or down, but remains in a fixed position as the load is moved.

8. What is accomplished by using a fixed pulley?

Figure 5.16–1. Fixed Pulley System

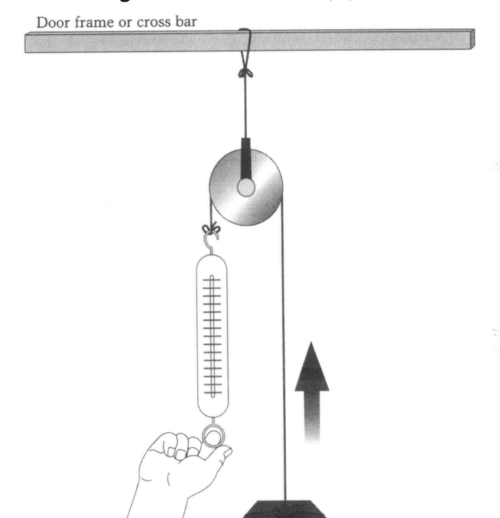

Door frame or cross bar

Teacher Information

When you use a fixed pulley, the load moves up or down, but the pulley itself is fastened to a stationary object and therefore remains in a fixed position.

A fixed pulley does not alter the amount of force required to lift an object, but it reverses the direction of the force; that is, as the force is applied in a downward direction the load is lifted in an upward direction.

The fixed pulley is a form of a turning first-class lever. (See Figure 5.16–2.) Think of the fulcrum as being at the center of the axle, the effort at one edge of the pulley wheel, and the load at the other edge. As with other first-class levers, the fulcrum is between the effort and the load. Curtains, drapes, and louvered blinds use fixed pulleys.

Integrating

Math

Science Process Skills

Observing, inferring, measuring, predicting, communicating, comparing and contrasting, formulating hypotheses, identifying and controlling variables, experimenting

Figure 5.16–2. Fixed Pulley System as a First-Class Lever

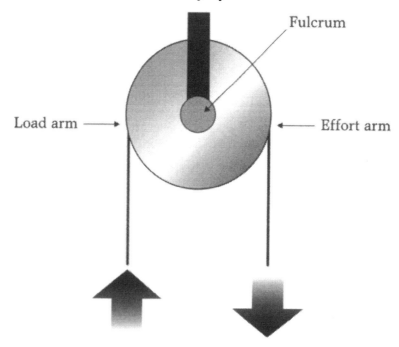

Measuring with a Fixed Pulley

Weight = _____ kg

a. Travel Direction

What direction does the load (books) move as the effort (spring balance) moves downward?

b. Travel Distance

How far does the effort move as the load moves 20 cm?

c. Force

How much force is required to lift the load?

What Is a Movable Pulley?

Materials Needed

- Pulley
- Crossbar
- Spring balance
- Meter stick (or yardstick)
- Cord or string
- Pencils
- "Measuring with a Movable Pulley" chart for each student
- Bundle of books (or other heavy object)
- Copies of the "Science Investigation Journaling Notes" for this activity for all students

Procedure

1. As you complete this activity, you will keep a record of what you do, just as scientists do. Obtain a copy of the form "Science Investigation Journaling Notes" from your teacher and write the information that is called for, including your name and the date.

2. For this activity, you will learn about movable pulleys. For item 1, the question is provided for you on the form.

3. Item 2 asks for what you already know about the question. If you have some ideas about movable pulleys and why we use them, write your ideas.

4. For item 3, write a statement of what a movable pulley is, according to what you already know, and that will be your hypothesis.

5. Now continue with the following instructions. Complete your Journaling Notes as you go. Steps 6 through 19 below will help you with the information you need to write on the form for items 4 and 5.

6. Use the "Measuring with a Movable Pulley" chart for recording your measurements in this activity.

7. Weigh the books with the spring balance and record the results.

8. Arrange your pulley, crossbar, spring balance, cord, and bundle of books, as shown in Figure 5.17–1, with one end of the cord attached to the crossbar.

Figure 5.17–1. Movable Pulley System

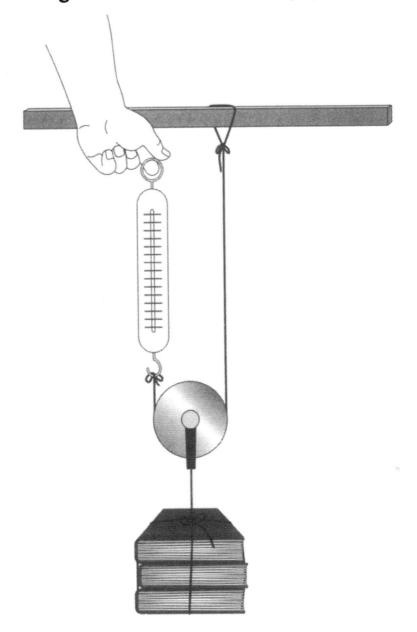

9. Lift the books by pulling up on the spring balance. Note the force indicated on the spring balance as you lift in a steady motion. Record this amount as the force required to lift the books using a movable pulley.

10. Compare the weight of the books with the amount of force required to lift the books using the movable pulley.

Simple Machines

11. Measure and record the distance traveled by the effort (spring balance) as the load (books) travels 20 cm.

12. Predict the distance the load will travel as the effort travels 30 cm.

13. Try it. Record the results and compare with your prediction.

14. Notice and record the direction traveled by load and effort as you lift the load.

15. Examine the information in your chart and consider these questions:

    a. Did the pulley decrease the amount of force needed to lift the books?

    b. Did the pulley cause the load to move a greater or lesser distance than the effort moved?

    c. Did the pulley cause the load to move in the opposite direction from that of the effort?

16. With the pulley fastened to the load, as it has been for this activity, and one end of the cord fastened to the crossbar, the pulley is called a movable pulley. The pulley itself moves up or down with the load.

17. What effect does a movable pulley have on the force required to lift a load?

18. What effect does a movable pulley have on the distance a load travels compared with the distance traveled by the effort?

19. What effect does a movable pulley have on the direction of the load compared with the direction of the effort?

20. Complete your "Science Investigation Journaling Notes." Are you ready to explain what a movable pulley is and how it helps us to do work? Discuss it with your group.

Teacher Information

A movable pulley is attached to the load and therefore moves up and down with the load.

When you use a movable pulley, the direction of travel does not change; the load travels in the same direction as the effort. However, the amount of force required to lift a load is less than the actual weight of the object. The travel distance of the load is less than that of the effort. Thus, the movable pulley offers an advantage as to force required, but it does so at the expense of travel distance and speed.

In computing the gain or loss from using a movable pulley, use the following as a guide:

a. The load will travel half the distance of the effort.
b. The force required to lift the load is half the actual weight of the object.

Note: The force required will be increased by whatever friction is involved as the pulley turns on the axle, the cord rubs against the sides of the groove in the pulley, and so forth.

Integrating

Math

Science Process Skills

Observing, inferring, measuring, predicting, communicating, comparing and contrasting, formulating hypotheses, identifying and controlling variables, experimenting

Measuring with a Movable Pulley

Weight of books = _____ kg

Force required to lift books = _____ kg

Travel Distance

What is the distance traveled by the effort (spring balance) as the load (books) moves 20 cm?

_____

What is the predicted distance the load would travel as the effort moves 30 cm?

_____

What is the actual distance the load travels as the effort moves 30 cm?

_____

Travel Direction

What is the direction traveled by the load and effort as you lift the load?

Load _____

Effort _____

Science Investigation

Journaling Notes for Activity 5.17

1. Question: *What is a movable pulley?*

2. What we already know:

_____

_____

3. Hypothesis:

_____

_____

4. Materials needed:

_____

_____

5. Procedure:

_____

_____

6. Observations/New information:

_____

_____

7. Conclusion:

_____

_____

What Happens When a Small Person Tugs on a Large Person?

Materials Needed

- Pulley
- Rope
- Two chairs with sturdy legs

Procedure

1. Anchor the pulley to Chair A, as seen Figure 5.18–1.

Figure 5.18–1. Two Chairs Tied Together, with Pulley Attached to Chair A

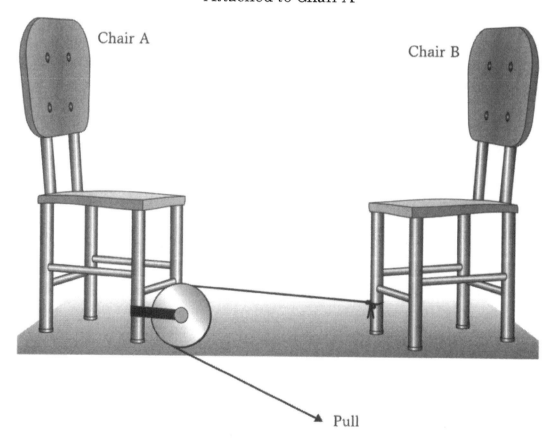

2. Thread one end of the rope through the pulley and tie the end of the rope to Chair B.

3. Have a large person sit on Chair A and a small person on Chair B.

4. When you pull on the other end of the rope, which chair will move?

5. After you make your prediction, pull on the rope.

6. What happened? Why?

7. Discuss your ideas with your group.

For Problem Solvers

Try to figure out a way to measure the force that is being exerted on each of the chairs. If you have more pulleys, figure out a way to make it still easier to pull the larger person sitting on a chair.

Teacher Information

Two ropes are pulling on Chair A, while only one rope is pulling on Chair B. The two ropes pulling on Chair A have equal tension. Twice as much force is exerted on Chair A as on Chair B.

If two spring scales are used in the system, one attaching the rope to Chair B, and the other is at the other end of the rope and the "pull" point, the difference in forces shown should be evident.

Integrating

Math, language arts

Science Process Skills

Observing, inferring, measuring, predicting, communicating, using space-time relationships, formulating hypotheses, identifying and controlling variables, experimenting

What Can Be Gained by Combining Fixed and Movable Pulleys?

Materials Needed

- Two pulleys
- Crossbar
- Spring balance
- Meter stick (or yardstick)
- Cord or heavy string
- Pencils
- "Measuring with a Combined Fixed and Movable Pulley" chart for each student
- Bundle of books (or other heavy object)

Procedure

1. Use the "Measuring with a Combined Fixed and Movable Pulley" chart for recording your measurements in this activity.
2. Weigh the books and record the weight in Part One.
3. Arrange the pulleys, crossbar, spring balance, cord, and bundle of books as illustrated in Figure 5.19–1. Notice that one pulley and one end of the cord are attached to the crossbar.
4. You now have a pulley system that includes a fixed pulley and a movable pulley.
5. From your previous experience, see whether you can predict the answers to the following questions. Record your predictions in Part One of the chart.
   a. Which direction will the load (books) move as you pull down at the effort position (spring balance)?
   b. Considering the actual weight of the books, how much force will be required to lift the load?
   c. How far will the load travel as the effort travels 40 cm?
6. After recording your predictions, test them by actual measurement. Record your measurements and compare them with your predictions.

7. Now record the following in Part Two:
   a. Change the number of books in your bundle, record the weight of your new load, and predict the amount of force necessary to lift it with the pulley system.
   b. Predict the travel distance of the effort as you lift the load 5 cm.
8. After recording your predictions, test them and record your actual measurements.

Figure 5.19–1. Pulley System with One Fixed and One Movable

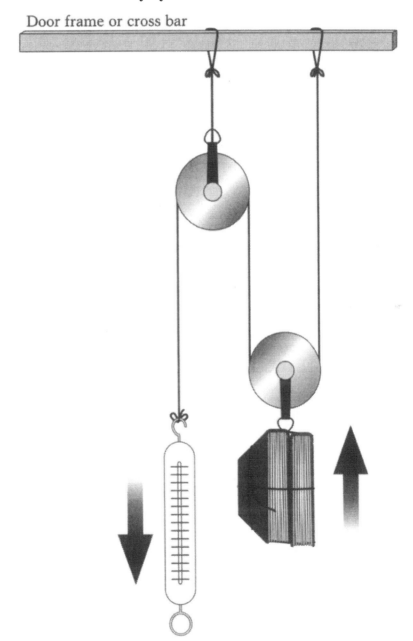

Door frame or cross bar

9. Were your predictions closer this time?

10. Can you think of any situation for which it would be helpful to combine a fixed pulley with a movable pulley?

For Problem Solvers

Try to find an opportunity to visit a crane while it is working at a construction site. Find as many simple machines as you can, including all three types of levers, fixed pulleys, movable pulleys, and combined pulleys.

Teacher Information

A combination of a fixed pulley and a movable pulley offers the advantages of both. A load can be lifted with half as much force as would be expected by considering the actual weight of the objects (because of the movable pulley) and the load can be moved upward by pulling downward (because of the fixed pulley).

This system, including a fixed pulley combined with a movable pulley, is called a block and tackle. It is used for lifting automobile motors, for raising and lowering scaffolds for painters, and for many other purposes.

Enrichment

Allow students to experiment with various combinations of pulleys and test the mechanical advantage of their creations.

Combinations of more than one fixed pulley and an equal number of movable pulleys multiply the mechanical advantage of the movable pulley and decrease efficiency in terms of travel distance. For instance, in a system involving a double fixed pulley and a double movable pulley, a load of 4 pounds would be lifted with approximately 1 pound of force at the effort position, but the effort must travel 4 cm (or inches) for each cm (or inch) the load is to be lifted.

If a strong overhead beam (such as a tree branch) is available, students would enjoy experimenting with their pulleys in lifting heavier objects, such as each other, the teacher, or several students at a time. *Caution:* Such an activity should be closely supervised to assure safety. Beware of possible broken ropes and falls. Vertical distance lifted should be limited to avoid risk of injury.

Integrating

Math, social studies

Science Process Skills

Observing, inferring, measuring, predicting, communicating, comparing and contrasting, formulating hypotheses, identifying and controlling variables, experimenting

Measuring with a Combined Fixed and Movable Pulley

Part One

Weight _____ kg

	Predicted Results	Measured Results
Load direction	_____	_____
Force required	_____	_____
Load distance	_____	_____

Part Two

Weight _____ kg

	Predicted Results	Measured Results
Force required	_____	_____
Effort distance	_____	_____

What Is an Inclined Plane?

Materials Needed

- Board at least 1.5 m (4.5 ft.) long
- Roller skate (or toy truck or small wagon)
- Cord or heavy string
- Pencils
- "Measuring with an Inclined Plane" chart for each student
- Spring balance
- Box of rocks (or books or other weights)
- Copies of the "Science Investigation Journaling Notes" for this activity for all students

Procedure

1. As you complete this activity, you will keep a record of what you do, just as scientists do. Obtain a copy of the form "Science Investigation Journaling Notes" from your teacher and write the information that is called for, including your name and the date.

2. For this activity, you will learn about inclined planes. For item 1, the question is provided for you on the form.

3. Item 2 asks for what you already know about the question. If you know something about inclined planes and why we use them, write your ideas.

4. For item 3, write a statement about what an inclined plane is, according to what you already know, and that will be your hypothesis.

5. Now continue with the following instructions. Complete your Journaling Notes as you go. Steps 6 through 20 below will help you with the information you need to write on the form for items 4 and 5.

6. Use the "Measuring with an Inclined Plane" chart for recording your measurements in this activity.

7. Using the string, attach the load to the roller skate.

8. Weigh the load, including the roller skate, and record its weight.

9. Place the board on a stairway, with one end at the bottom of the stairs and the other end resting on the fourth step.

10. Attach the spring balance to the skate and pull the load up the inclined plane (board), as shown in Figure 5.20–1. As you pull steadily, notice the force indicated by the needle and record the results.

11. Compare the weight of the load with the force needed to pull the load up the inclined plane.

12. As you pulled the load up the inclined plane, was more force or less force required than the actual weight of the load? Record your answer.

13. Move the top of the board down to the first step, as shown in Figure 5.20-2.

14. Judging from your first experience, how much force do you think will be required to pull the load up the slope? Record your prediction.

15. After recording your prediction, attach the spring balance to the load and try it.

16. Record the results. How close was your prediction?

17. Now predict the force required with the board on the second step. Try it after recording your prediction.

Figure 5.20-1. Roller Skate with Load on Inclined Plane

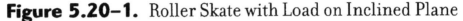

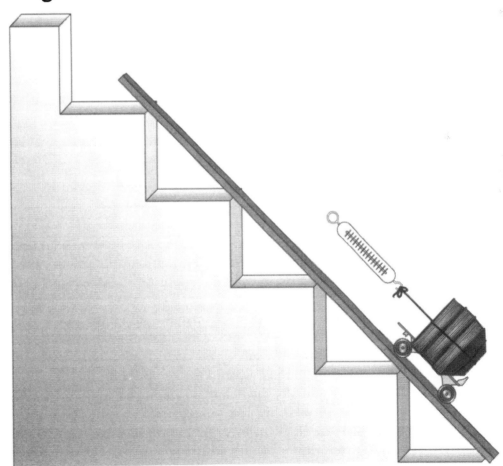

Figure 5.20–2. System in New Position

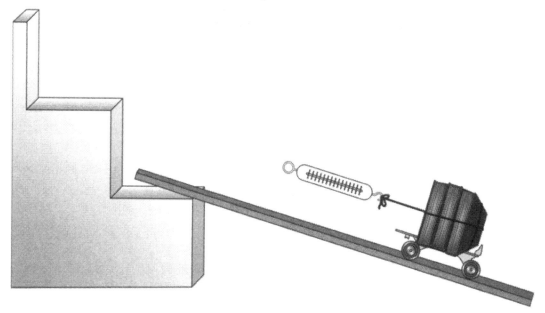

18. Was your prediction closer this time?

19. If your load weighed 100 pounds, how much force would you need to apply to push it up the inclined plane with the top of the inclined plane resting on the first step?

20. Think of some ways inclined planes would be useful. Write down two of them and show them to your teacher.

21. Complete your "Science Investigation Journaling Notes." Are you ready to explain what an inclined plane is? Discuss it with your group.

Teacher Information

An inclined plane is a sloped surface. It provides a mechanical advantage of force. We can move a load up an inclined plane with less force than would be indicated by the actual weight of the object (provided the friction isn't too great).

As with the use of any other machine, the total work required is not reduced but only redistributed. The advantage gained in force is sacrificed in distance and speed.

The ideal mechanical advantage is computed by dividing the length of the inclined plane by the height, as in Figure 5.20–3. Friction will enter in, according to whether the object is forced to slide up the inclined plane, moved on rollers, or by other means. The actual mechanical advantage can be determined by dividing the weight of the object by the force

Hands-On Physical Science Activities

required to move the object. Friction can be helpful in this application, because it helps to keep the load from slipping backward.

The weight of the object is twice the force required to move the object up the ramp (if friction could be eliminated), but we must move the object twice as far as if we lifted it straight up (eight units instead of four units).

Application of the inclined plane is illustrated as barrels of oil are rolled up a ramp, as a car drives up a mountain on a winding road, and as a person walks up a stairway or up the sloping floor of a theater.

Note: If it is inconvenient for students to use a stairway for their work with the inclined plane, have them support the end of their board on a stack of books or some other sturdy, adjustable support.

Integrating

Math, social studies

Science Process Skills

Observing, inferring, measuring, predicting, communicating, comparing and contrasting, formulating hypotheses, identifying and controlling variables, experimenting

Figure 5.20–3. Inclined Plane System

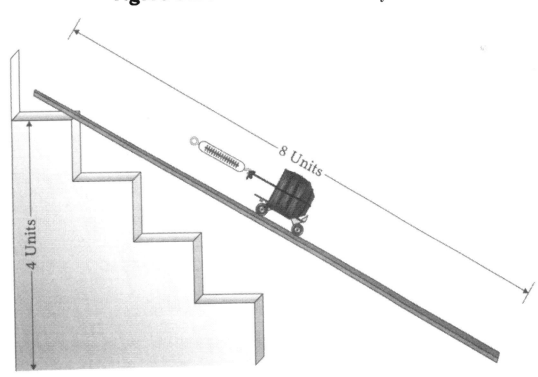

Measuring with an Inclined Plane

Weight = _____ kg

Position of Upper End	Position of Lower End	Force
4th step	bottom	more/less (circle one)
1st step	bottom	Prediction _____ Actual _____
2nd step	bottom	Prediction _____ Actual _____
1st step		Predicted force to move 100 pounds _____

Uses for inclined planes:

1. _____

2. _____

Science Investigation
Journaling Notes for Activity 5.20

1. Question: *What is an inclined plane?*

2. What we already know:

_____

_____

3. Hypothesis:

_____

_____

4. Materials needed:

_____

_____

5. Procedure:

_____

_____

6. Observations/New information:

_____

_____

7. Conclusion:

_____

_____

What Is a Wedge?

Materials Needed

- Wedge
- Stack of books
- Board

Procedure

1. Stack the books up on one end of the board.
2. Place your fingers under the end of the board near the books and lift it up about 3 to 8 cm. Notice the force required to lift the books.
3. Now place the tip of the wedge under the same end of the board, as shown in Figure 5.21-1.
4. Tap the wedge with your foot, forcing it under the end of the board.
5. What is happening to the load of books?
6. Can you tell whether the force required to drive the wedge under the board is greater or less than the force required to lift the load directly?
7. Why does the wedge so strongly resist being forced under the board?

Figure 5.21–1. Board, Books, and Wedge

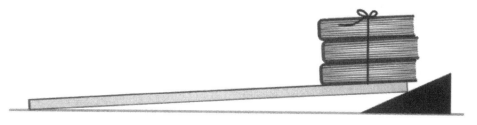

For Problem Solvers

Make a list of all the examples of wedges that you can find.

Teacher Information

Two wedges can be made by sawing a block of wood in half diagonally, as seen in Figure 5.21–2. Note that each wedge looks like an inclined plane. It differs from the inclined plane only in its application. When it

Hands-On Physical Science Activities

is used as an inclined plane, an object (load) moves up the incline. When it is used as a wedge, the inclined plane moves into, or under, the object.

We gain in force and also change the direction of the force by using a wedge.

The longer or thinner the wedge, the greater the gain in force and the greater the loss in distance, that is, the farther the wedge must go under an object to lift it a given amount.

The maximum distance a load can be moved by a wedge is the thickness of the larger end of the wedge.

The ideal mechanical advantage of the wedge can be computed by dividing the length of the wedge by the thickness of the larger end. However, because of the great amount of friction that is usually involved, the actual mechanical advantage is almost always significantly less than the ideal. Friction, however, is often helpful when using a wedge because it keeps the wedge from slipping out.

Applications of the wedge are the ax, chisel, tip of a nail or pin, knife blade, wood-splitter's wedge, and so forth.

Enrichment

If a short log, a sledge hammer, and a wood-splitting wedge could be acquired, a demonstration by an adult of the use of the wedge in actually splitting the log would provide an excellent experience in seeing the usefulness of the wedge. *Caution:* Splitting wood may splinter, so use care and be sure children stand back a safe distance.

Integrating

Math, social studies

Science Process Skills

Observing, inferring, measuring, predicting, communicating, comparing and contrasting, formulating hypotheses, identifying and controlling variables, experimenting

Figure 5.21–2. Cut in Block of Wood to Make Wedges

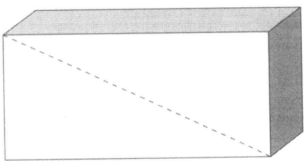

What Is a Screw?

Materials Needed

- Pencils
- Sheets of white paper
- Scissors
- Black markers
- Large wood screws
- Rulers

Procedure

1. Cut a triangle from a sheet of white paper by cutting diagonally, from corner to corner. (See Figure 5.22–1.)

2. Using a marker, make a heavy black line along the hypotenuse (the longest side), as in Figure 5.22–2.

3. Hold the triangle upright on top of your table or desk. Which of the simple machines we have already studied does the triangle now look like?

4. Beginning with the long side, wrap the paper triangle around a pencil. (See Figure 5.22–3.)

5. Hold the wrapped pencil side by side with a large wood screw.

6. Does the heavy line of the triangle resemble the threads of the screw?

7. How would you say the threads of a screw compare to an inclined plane? Think about how the triangle with its black edge looked before you wrapped it around the pencil.

Figure 5.22–1. Rectangular Paper with Diagonal Cut

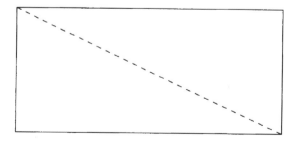

Hands-On Physical Science Activities

Figure 5.22–2. Triangle with Darkened Hypotenuse

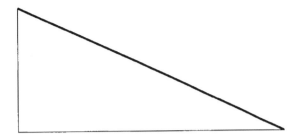

Figure 5.22–3. Triangle Being Wrapped Around Pencil

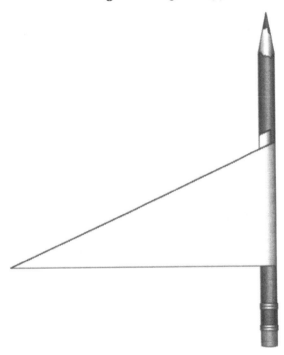

Teacher Information

The screw is a form of inclined plane that winds around in a spiral. The spiral-shaped ridge around the screw is called the thread.

Integrating

Math, social studies

Science Process Skills

Observing, inferring, measuring, predicting, communicating, comparing and contrasting, formulating hypotheses, identifying and controlling variables, experimenting

What Kind of Simple Machine Is the Screwdriver?

Materials Needed

- Wood screws
- Boards (pine or other soft wood)
- Screwdrivers

Procedure

1. Try to push a screw into a board with your fingers.
2. Place the tip of a screwdriver on the head of the screw and try to push the screw into the board without turning the screwdriver.
3. Did the screw go any farther than when you pushed with your hand?
4. Now turn the screwdriver as you press down hard with it on the head of the screw.
5. Can you get the screw into the board this time?
6. Notice how far the screwdriver handle travels as the screw turns around once.
7. How far does the screw move into the board as the screwdriver handle turns around?
8. What can you say about the gain or loss in force and in distance as you turn the screw into the board?
9. Make a list of ways people make use of the screw.
10. What kind of machine is the screwdriver?

Teacher Information

The screw offers a gain in force, but at the expense of distance and speed. The handle of the screwdriver travels much faster and farther in a circular direction than the distance the screw moves into the board.

The screw also changes the direction of the effort from a turning motion to a pulling motion.

The ideal mechanical advantage of the screw is computed by dividing the circumference of the screwdriver handle (distance it travels in one complete turn) by the pitch (distance between two threads) of the screw. However, the actual mechanical advantage of the screw is greatly reduced by friction.

Friction helps by keeping the screw from turning backward or pulling out. The loss of force due to friction is made up by using another machine, the screwdriver. The screwdriver is a form of the wheel and axle.

The principle of the screw is applied in the use of the base of a light bulb, a bolt, the adjusting wheel of a pipe wrench, caps on jars and bottles, the piano stool, and many other applications.

Integrating

Math, social studies

Science Process Skills

Observing, inferring, measuring, predicting, communicating, comparing and contrasting, formulating hypotheses, identifying and controlling variables, experimenting

What Kind of Simple Machine Is This?

Materials Needed

Variety of simple and complex machines, as available, such as:

- Ball bat
- Bumper jack
- Eggbeater
- Food grinder
- Pencil sharpener
- Rake
- Screw jack
- Tongs
- Fishing pole
- Broom
- Can opener (any kind available)
- Flour sifter
- Hand drill
- Pepper mill
- Scissors
- Shovel
- Tweezers
- Golf club
- Paper and pencils

Procedure

1. Examine each of the devices. First, classify them into two groups, as simple machines or complex machines.

2. Next, classify all of the simple machines according to the type of simple machines they are. Make a group of levers, wedges, and so forth. Then group the levers as first-class, second-class, and third-class.

3. Now look at those you put in the group of complex machines. For each of these, make a list of the simple machines you see in them.

For Problem Solvers

Make a list of all the machines you have used today. For each one, name all the simple machines that are in it.

Examine a toy that has internal working parts. Think carefully about how it works and draw what you think is inside. What simple machines are there, according to your drawing? Do another one, and compare your ideas for each one with someone else who is doing the activity, or with your teacher or someone in your family.

Teacher Information

This activity will reveal what your students have really learned about simple machines. Try to allow them time to pursue the ideas suggested in "For Problem Solvers" as long as their interest lasts. Perhaps they can continue at home if necessary.

Integrating

Math, social studies

Science Process Skills

Observing, inferring, measuring, predicting, communicating, comparing and contrasting, formulating hypotheses, identifying and controlling variables, experimenting

Can You Solve This Simple Machines Word Search?

Try to find the following Simple Machines terms in the grid below. They could appear in horizontal (left to right), vertical (up or down), or diagonal (upward or downward) position.

machine	work	energy
lever	wheel	axle
wedge	advantage	pulley
incline	force	mechanical
fulcrum	load	effort
friction		

```
E  F  F  O  R  T  M  F  O  R  C  E
N  B  V  C  X  Z  F  U  A  S  D  N
P  L  K  W  J  I  R  L  H  G  F  E
M  E  C  H  A  N  I  C  A  L  T  R
O  I  U  E  K  C  R  D  O  Y  G
M  W  E  E  R  L  T  U  V  A  R  Y
M  A  N  L  O  I  I  M  A  D  P  B
S  A  C  Z  W  N  O  X  N  C  U  V
D  F  G  H  J  E  N  K  T  L  L  P
R  T  Y  U  I  I  D  O  A  X  L  E
E  R  T  Y  U  N  I  G  G  K  E  M
V  G  B  L  E  V  E  R  E  L  Y  J
```

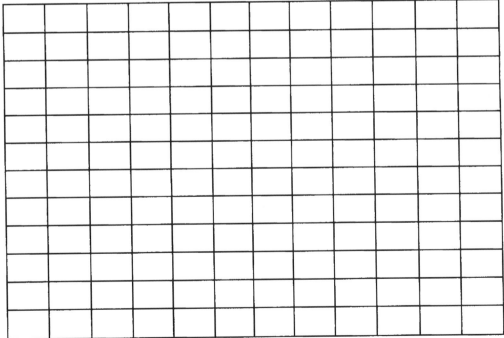

Can You Create a Simple Machines Word Search of Your Own?

Write your Simple Machines words in the grid below. Arrange them in the grid so they appear in horizontal (left to right), vertical (up or down), or diagonal (upward or downward) position. Then fill in the blank boxes with other letters. Trade your Word Search with someone else who has created one of his or her own, and see whether you can solve the new puzzle.

_____   _____   _____

_____   _____   _____

_____   _____   _____

_____   _____   _____

_____   _____   _____

Simple Machines

Answer Key for
Simple Machines Word Search

```
E  F  F  O  R  T  M  F  O  R  C  E
N  B  V  C  X  Z  F  U  A  S  D  N
P  L  K  W  J  I  R  L  H  G  F  E
M  E  C  H  A  N  I  C  A  L  T  R
O  I  U  E  K  C  C  R  D  O  Y  G
M  W  E  E  R  L  T  U  V  A  R  Y
M  A  N  L  O  I  I  M  A  D  P  B
S  A  C  Z  W  N  O  X  N  C  U  V
D  F  G  H  J  E  N  K  T  L  L  P
R  T  Y  U  I  I  D  O  A  X  L  E
E  R  T  Y  U  N  I  G  G  K  E  M
V  G  B  L  E  V  E  R  E  L  Y  J
```

Do You Recall?

Section Five: Simple Machines

1. Name two things friction does that reduce the efficiency of machines.

2. What two things are accomplished by using a first-class lever?

3. How is a second-class lever different from a first-class lever?

4. What is the major advantage of the third-class lever?

Do You Recall? *(Cont'd.)*

5. Name one or more examples of the wheel and axle.

6. What advantage is provided by the fixed pulley?

7. What is the advantage of the movable pulley?

8. What is the difference between the inclined plane and the wedge?

9. When an inclined plane spirals around and around, what is it called?

Answer Key for Do You Recall?

Section Five: Simple Machines

Answer	Related Activities
1. Produces heat and causes resistance	5.1–5.4
2. a. Change the direction the load is moved	
b. Reduce the amount of force necessary to move the load (if the fulcrum is closer to the load than to the effort)	5.5–5.9
3. a. The load moves in the same direction as the effort.	
b. The load is between the fulcrum and the effort.	5.10, 5.11
4. The load moves faster and farther than does the effort.	5.12, 5.13
5. Steering wheel, screwdriver, doorknob (the wheel is attached to the axle, so one turns the other)	5.14, 5.15
6. Only that it changes the direction the load is pulled	5.16
7. The amount of effort required to move the load is reduced.	5.17–5.19
8. The load is moved up the inclined plane, but the wedge is forced under or into the load.	5.20, 5.21
9. A screw	5.22

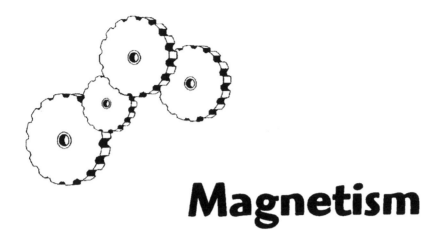

Magnetism

To the Teacher

A study of magnetism is often a very helpful beginning point for introducing structured inquiry/discovery activities. Be certain to warn the children that magnets can break or lose their magnetism if dropped or hit together. In the case of U-shaped magnets, a piece of soft iron called a keeper should be placed across the ends.

This material is nongraded. However, some activities may seem more appropriate to certain age levels. Each teacher should feel free to reorganize and eliminate materials according to the needs of students, keeping in mind the level of psychological development that may influence the understanding of certain concepts. Using the inquiry approach, complete with concrete materials, is very important with this section.

Many of the activities in this section seem to be most effective when presented to individuals, small groups, or teams. If your classroom organization permits, consider placing the materials in a science learning center, with time provided for children to move through the sequence at their own rate. Classroom discussions to reinforce the concepts should follow. If classroom demonstrations are used initially, children should perform the activities, and then the materials should be left on the science table for children to explore individually.

Children are fascinated by magnets. They seem almost like magic. There is a natural desire on the part of many children to explore and share their discoveries with friends and others outside the classroom. Most magnets are fairly expensive, but many school-supply outlets sell small magnets (for notices on bulletin boards, refrigerator doors, and so on) at a reasonable cost. If possible, try to get a number of these small magnets for out-of-school activities.

One excellent source of magnets is audio speakers. Every speaker contains a magnet. They come in many different sizes and strengths, and they are generally fairly easy to remove with a hammer and screwdriver. They are usually riveted in place; some are held with bolts and nuts. Check with a local shop that installs car stereos; they usually throw the old speakers away. Occasionally, a speaker system in a school or other large building will be replaced, and you might be able to get some dandies.

Bar magnets can be given new life by stroking them lengthwise several times across one pole of a powerful magnet. Stroke in one direction only. If poles are reversed, stroke in the opposite direction or use the other pole of the large magnet.

Caution: Audiotapes, videotapes, and computer disks are recorded magnetically. If they come near a strong magnet, they may be erased. Spring-operated watch mechanisms may also become magnetized in a strong magnetic field, although they are rarely seen these days.

The following activities are designed as discovery activities that students can usually perform quite independently. You are encouraged to provide students (usually in small groups) with the materials listed and a copy of the activity from the beginning through the "Procedure." The section titled "Teacher Information" is not intended for student use, but rather to assist with discussion following the hands-on activity, as students share their observations. Discussion of conceptual information prior to completing the hands-on activity can interfere with the discovery process.

Regarding the Early Grades

With verbal instructions and slight modifications, many of these activities can be used with kindergarten, first-grade, and second-grade students. In some activities, steps that involve procedures that go beyond the level of the child can simply be omitted and yet offer the child an experience that plants the seed for a concept that will germinate and grow later on.

Teachers of the early grades will probably choose to bypass many of the "For Problem Solvers" sections. That's okay. These sections are provided

Hands-On Physical Science Activities

for those who are especially motivated and want to go beyond the investigation provided by the activity outlined. Use the outlined activities, and enjoy worthwhile learning experiences together with your young students. Also consider, however, that many of the "For Problem Solvers" sections can be used appropriately with young children as group activities or as demonstrations, still giving students the advantage of an exposure to the experience and laying groundwork for connections that will be made later on.

Correlation with National Standards

The following elements of the National Standards are reflected in the activities of this section.

K-4   Content Standard A: Science as Inquiry

As a result of activities in grades K-4, all students should develop:

1. Abilities necessary to do scientific inquiry
2. Understanding about scientific inquiry

K-4   Content Standard B: Physical Science

As a result of activities in grades K-4, all students should develop understanding of

1. Properties of objects and materials
2. Position and motion of objects
3. Light, heat, electricity, and magnetism

5-8   Content Standard A: Science as Inquiry

As a result of activities in grades 5-8, all students should develop:

1. Abilities necessary to do scientific inquiry
2. Understanding about scientific inquiry

5-8   Content Standard B: Physical Science

As a result of activities in grades 5-8, all students should develop understanding of

1. Properties and changes of properties in matter
2. Motions and forces
3. Transfer of energy

Which Rock Is Different?

Materials Needed

- Several similar rocks per small group
- One lodestone per small group
- Paper clips

Procedure

1. One of the rocks can do something the others cannot.
2. Can you find it?
3. What can you say about it?

For Problem Solvers

Look up the word *lodestone* in an encyclopedia, the Internet, or other reference book and learn all you can about this interesting rock. How is it different from other rocks? How is it different from other magnets?

Teacher Information

The children may choose a rock other than the magnetic lodestone. If this occurs and their reasons for the choice are logical, their answers should be accepted, as the process of inquiry is your objective. However, since it is assumed that most children have had some experience with magnets, a paper clip or other magnetic material on the table may assist in the discovery.

Lodestone is a particular type of iron ore that occurs naturally and has properties of magnets. Lodestones may be purchased from a science-supply house at a nominal cost. Iron ore that is magnetic (attracted to a magnet) but does not behave as a magnet is called magnetite. Lodestone behaves as a magnet and has magnetic poles. Most people use the two terms interchangeably, in reference to the rock that behaves as a magnet.

Integrating

Reading

Science Process Skills

Observing, inferring, comparing and contrasting, researching

What Do Magnets Look Like?

Materials Needed

- A large collection of magnets of different sizes, shapes, materials, and colors
- Magnetic materials, such as paper clips, for testing

Procedure

1. What do magnets look like?
2. Use the materials provided by your teacher to answer the question.

For Problem Solvers

Make a list or make drawings of all of the shapes of magnets you have in the classroom. Then explore elsewhere, at home, at your parents' workplace if you can get permission, and wherever you can find magnets. Add to your list all that you find. Ask your friends if they know of still other types of magnets. Add these to your list, too.

Make another list of all the places and ways you can find that magnets are being used. Do you have any on your refrigerator? On your cupboard door latches? Where else can you find magnets being used? Ask your family and friends. Ask a mechanic whether any magnets are used in automobiles or other equipment that he or she works with. Ask about electric motors. Do you have a metal salvage yard nearby? Ask about the use of magnets there.

If you have science books about magnets, you might find still more ideas there.

Teacher Information

The answer to the question is simple. The appearance of magnets varies a lot. You cannot tell if something is a magnet by its appearance. If possible, obtain "cow magnets" from feed and grain stores. These magnets are put in the stomachs of cattle to collect bits of metal that have been eaten.

Children often have the idea that magnetism has something to do with the shape of the magnets. This activity will help children see that shape is not directly related to the property of magnetism.

Students might have unusual magnets of their own that they could bring to add variety to the shapes of magnets used.

Caution: Strong magnets can pinch little fingers; close supervision is necessary.

Integrating

Language arts, social studies

Science Process Skills

Observing, inferring, classifying, predicting, communicating, comparing and contrasting, researching

How Do Magnets Get Their Names?

Materials Needed

- Variety of magnets

Procedure

1. Magnets often get their names from their shapes or from what they do. See whether you can find magnets that might have the following names:

    bar magnet

    cylindrical magnet

    disk (or disc) magnet

    U-shaped magnet

    horseshoe magnet

    cow magnet

2. Can you name any others?

Teacher Information

This activity may have some value in helping children learn new names for certain shapes. It will also provide a basic vocabulary for further study of magnets. Since magnets are so much a part of our everyday life, your students may bring in many new and unusual magnets to add to the collection.

Integrating

Language arts

Science Process Skills

Observing, inferring, classifying, comparing and contrasting

Where Did the First Metal Magnet Come From?

Materials Needed

- Nonmagnetized needles (with point broken off)
- Lodestones
- Paper clips

Procedure

1. Is the needle a magnet?
2. Test it by trying to pick up a paper clip or some other small object.
3. Rub it twenty times in the same direction with the magnetic rock.
4. Test your needle again. Is it a magnet?
5. What can you say about this?

For Problem Solvers

Check reference books and the Internet for the words *magnet* and *magnetism*. Try to find out who used magnets first and what they used them for. What kind of magnets did they use? How did people make the first magnets that were made by people?

Teacher Information

Magnetism has been known for centuries. References to this "magical" property occur in Chinese and Greek mythology. This activity might provide opportunities for creative writing about the discovery of magnetism. The Internet and/or your encyclopedia can provide information about the history of lodestones.

The point of the needle can be easily broken off with a pair of pliers.

Integrating

Reading, language arts, social studies

Science Process Skills

Observing, inferring, researching

What Materials Will a Magnet Pick Up?

(Take home and do with family and friends.)

Materials Needed

- A tray of magnetic and nonmagnetic materials, such as a Canadian nickel, a U.S. nickel, and brass and steel safety pins (be sure to include gold-colored items, silver-colored items, and items of other colors)
- Magnets
- Paper and pencils
- "Science Investigations Journaling Notes" for this activity for each student

Procedure

1. As you complete this activity, you will keep record of what you do, just as scientists do. Receive a copy of the form "Science Investigation Journaling Notes" at the end of this activity, and write the information that is called for, including your name and the date.

2. For this activity you will learn about what materials a magnet will pick up. For item 1, the question is provided for you on the form.

3. Item 2 asks for what you already know about the question. If you have some ideas about what magnets do, write your ideas.

4. For item 3, write a statement of what materials a magnet will pick up, according to what you already know, and that will be your hypothesis.

5. Now continue with the following instructions. Complete your Journaling Notes as you go. Steps 6 through 8 below will help you with the information you need to write on the form for items 4 and 5.

6. Test the objects you have, and find out which ones magnets can pick up or attract.

7. Make a list of these.

8. Make another list of things magnets will not attract.

9. Complete your "Science Investigation Journaling Notes." Are you ready to explain which materials magnets will pick up and which ones they won't? Discuss it with your group.

For Problem Solvers

Now that you have tested several small items to find out whether or not they are magnetic, take a magnet and test a lot of other things. Each time you think of a new material to check, make a prediction first, then check it out. Make a list of the materials that surprised you.

Classify your list of materials by making separate lists of things that are magnetic and things that are not magnetic.

Teacher Information

Children who have not yet acquired writing skills can make piles of magnetic and nonmagnetic materials. For older children, writing lists of their findings is a good exercise.

As students experiment with the materials, it is hoped that they will become motivated to seek the answer to the title question through the use of reference materials. The Internet, an appropriate video, library book, encyclopedia, or textbook can provide an explanation.

Materials that magnets will pick up are generally those made of iron, nickel, steel, and cobalt. These are called the strong magnetic materials.

Caution: Strong magnets can pinch little fingers; close supervision is necessary.

Integrating

Math, language arts

Science Process Skills

Observing, inferring, classifying, predicting, communicating, formulating hypotheses, researching

Science Investigation

Journaling Notes for Activity 6.5

1. Question: *What materials will a magnet pick up?*

2. What we already know:

_____

_____

3. Hypothesis:

_____

_____

4. Materials needed:

_____

_____

5. Procedure:

_____

_____

6. Observations/New information:

_____

_____

7. Conclusion:

_____

_____

Magnetism

Through What Substances Can Magnetism Pass?

Materials Needed

- Paper clip suspended from a string toward a large mounted magnet
- Variety of magnetic and nonmagnetic materials, such as a piece of plastic, wood, aluminum foil, paper, iron lid

Procedure

1. See Figure 6.6-1 for setting up this activity. What keeps the paper clip up?
2. Place different materials between the paper clip and the magnet.
3. What happened?
4. Discuss your observations with your group.

For Problem Solvers

Find other materials to test. Can you find any material that is not magnetic and that will block the magnetic force and cause the paper clip to fall?

Teacher Information

Use a string to suspend a strong magnet from a mount or anchored ruler, as shown in Figure 6.6-1. Place a paper clip held by a thread below the magnet, leaving a space between the magnet and the clip. The string can be attached to the floor with tape, or can be anchored by a book.

Caution: Strong magnets can pinch and injure little fingers; close supervision is necessary.

Nonmagnetic materials will pass between the clip and magnet without disturbing the magnetic field. Items passed between must not touch the paper clip or the magnet. When the iron lid comes near the gap, the magnet will be attracted to it and the clip will fall. In a later class discussion, it should be demonstrated that magnetic materials will disturb or cut a magnetic field. However, substances made of a magnetic material (iron, nickel, steel, or cobalt) may be made into temporary magnets and behave like magnets while in the magnetic field. Thus, the iron lid alone may not pick up paper clips, but when it comes near or touches a

Figure 6.6–1. Setup for Activity 6.6

strong magnet, it will temporarily become a magnet while in contact with the magnet. This is called *induction.* The extent of time the magnetic material retains the properties of a magnet after being removed from the magnet is related to its hardness (how tightly the molecules are packed together) and the specific ingredients used in the manufacture of the steel. A steel needle will retain its magnetism for a long time. A soft iron lid or paper clip will lose its magnetism rapidly when removed from the magnetic field.

There is no insulator to magnetism, but students will enjoy trying to find one.

Integrating
Language arts

Science Process Skills
Observing, inferring, classifying, predicting, communicating, researching

Which Magnet Is Strongest?

Materials Needed
- Variety of magnets
- Paper clips
- Rulers
- Spring scales

Procedure
1. Which magnet is strongest?
2. Use the materials provided to answer the question.
3. Discuss your observations and ideas with your group.

For Problem Solvers
Find some information about new types of magnets. Find out whether magnets have to be large in order to be strong.

Teacher Information
Often children will begin by putting two magnets together to see which one seems to "pull" harder. You may need to point out that this will not show which one is pulling harder. To help in the investigation, a spring scale, ruler, and paper clip are provided, but the children should not be given direction on how to use them. The purpose of the activity is to encourage creative inquiry. Success should be evaluated on the amount of creative exploration children undertake.

Younger children will often choose the largest magnet as the strongest. Teachers of young children should keep in mind the work of Piaget with pre-operational thinking in utilizing this activity. It's very natural to assume that a larger magnet is stronger than a smaller magnet.

Caution: Strong magnets can pinch little fingers; close supervision is necessary. Today's magnets include very strong magnets that are deceptively small.

Integrating
Math, language arts

Science Process Skills
Observing, inferring, classifying, predicting, communicating, comparing and contrasting, researching

What Part of a Magnet Has the Strongest Pull?

Materials Needed

- Bar magnets
- Paper clips

Procedure

1. See how many paper clips you can make stick to different parts of the magnet.
2. What happened?
3. Discuss your findings with your group.

Teacher Information

Magnets are strongest at the poles. Many more paper clips will be attracted at and near the poles than between the poles of the magnet. Few, if any, will cling to it between the poles. If there seems to be a point of magnetic strength along the length of a bar magnet where several paper clips cluster, this is still true, but the magnet has an additional set of poles. This can happen by inadvertently (or deliberately) placing the bar magnet in contact with the poles of a powerful magnet. If this should occur, the poles of the bar magnet can be repolarized by dragging it lengthwise several times across one pole of a powerful magnet. If the poles of the bar magnet are marked as N and S, you need to check it to see that the actual polarity matches the marked polarity. If the poles are reversed, simply repeat the treatment, but *turn the bar magnet end for end* and drag it across the *same pole* of the powerful magnet as before. Either reversing the direction of drag *or* using the opposite pole of the powerful magnet will reverse the polarity of the bar magnet.

Many of the bar magnets commonly used in classrooms are not very strong and are quite vulnerable to polarity changes when placed in contact with other magnets. They also tend to weaken over time. Bar magnets that seem to have lost their strength do not need to be discarded, but can be effectively rejuvenated by the process described.

The above activity will also illustrate the theory of induction. While the paper clips are touching the magnet, they temporarily become magnetized and several will "stick" to one another. When the clips are removed from the magnet, they rapidly lose their ability to attract each other.

Integrating

Math

Science Process Skills

Observing, inferring, measuring, predicting, communicating, formulating hypotheses, identifying and controlling variables, experimenting

What Is a Special Property of Magnetism?

(Take home and do with family and friends.)

Materials Needed

- Two bar magnets
- Thread or light string
- Piece of wire, somewhat stiff but easily bent, about 15 cm (6 in.) long
- Pliers

Procedure

1. Make a cradle for the magnet, something like the one shown in Figure 6.9–1.

2. Tie the thread to the top of the cradle and rest one bar magnet in the cradle. (See Figure 6.9–2.)

Figure 6.9–1. Cradle for Magnet

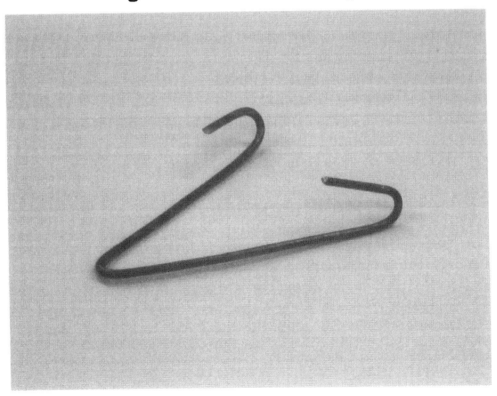

Figure 6.9–2. Magnet in Cradle

3. Hold the bar magnet in the air by the thread so it can turn freely.

4. Bring one end of the other bar magnet near the one that is suspended from the thread.

5. What happened?

6. Turn the bar magnet around and bring the other end near the one suspended from the thread.

7. Discuss your observations with others.

Teacher Information

Be sure the bars are correctly magnetized and that each has an "N" and an "S" marked on opposite ends. Like poles will repel each other, while unlike poles will attract one another.

The poles of the bar magnet can be reversed by drawing the bar magnet several times across one of the poles of a powerful magnet. The direction in which the bar magnet is drawn across the pole of the larger magnet will determine the polarity of the bar magnet.

The ability to attract and repel each other is a special characteristic by which magnets can be identified.

Caution: Strong magnets can pinch little fingers; close supervision is necessary.

Integrating

Math, language arts

Science Process Skills

Observing, inferring, predicting, communicating, experimenting

What Happens When a Magnet Can Turn Freely?

(Take home and do with family and friends.)

Materials Needed

- Bar magnet
- Thread or light string
- Piece of wire, somewhat stiff but easily bent, about 15 cm (6 in.) long
- Pliers

Procedure

1. Make a cradle for the magnet, something like the one shown in Figure 6.10–1.

2. Tie the thread to the top of the cradle and rest the bar magnet in the cradle. (See Figure 6.10–2.)

3. Hang the bar magnet in the air by the thread so it can turn freely. Tie it to a support and be sure there are no iron chair legs or other magnetic objects nearby.

Figure 6.10–1. Cradle for Magnet

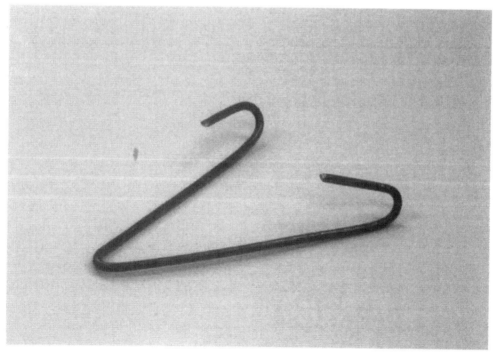

Hands-On Physical Science Activities

Figure 6.10–2. Magnet in Cradle

4. If the magnet spins, give it time to completely stop turning.
5. Observe the position in which it stops.
6. Do this several times.
7. What happened?
8. What can you say about this? Discuss your observations and ideas.

Teacher Information

If the bar magnet is correctly magnetized, it should stop each time with the N end pointing toward magnetic north. Calling the N end of the magnet north has come about through common usage, as people knew about magnets and their behavior long before they understood them scientifically. For clarification, since opposite poles attract, we should call the pole marked N the "north-seeking pole."

Integrating

Social studies

Science Process Skills

Observing, inferring, predicting, communicating

Magnetism 387

What Is a Compass?

Materials Needed

- Bar magnets
- Sensitive compasses
- Paper clips

Procedure

1. Bring the bar magnet near the compass.
2. What happens to the compass needle?
3. What can you say about this?
4. Now try the same thing with the paper clip. What happened this time?
5. Discuss your observations and ideas with your group.

Teacher Information

A compass is a freely suspended bar magnet. If a bar magnet is brought near, it will respond to the magnet.

The children may also bring other known magnetic materials near the compass to further reinforce this idea. Understanding that a compass is a magnet is important for further investigations in this unit.

Caution: Very strong magnets may damage a compass by pulling the needle off its delicate support.

Integrating

Language arts

Science Process Skills

Observing, inferring, predicting, communicating

How Can You Make a Compass?

Materials Needed

- Plastic, aluminum, or glass bowls
- Nonmagnetized needles (with the points broken off)
- Small pieces of plastic foam (or wood, cardboard, or anything that will float)
- Compasses
- Water
- Bar magnets

Procedure

1. Is the needle a magnet? Don't guess. Devise a way to find out.
2. Stroke the needle thirty times across the bar magnet in the same direction. Is the needle a magnet now?
3. Fill the bowl partly full of water.
4. Stick the needle through the plastic foam and float it in the pan of water, as shown in Figure 6.12–1.
5. Point the needle in different directions, then allow it to settle.

Figure 6.12–1. Needle in Floating Plastic Foam

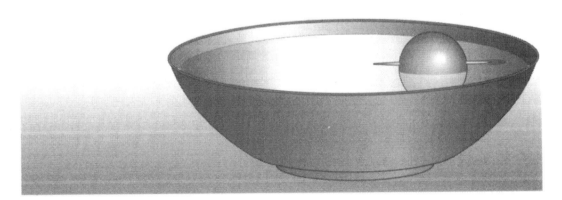

6. What happened?

7. Can you see a relationship to the compass?

8. What can you say about this?

For Problem Solvers

If you were lost in the woods and you didn't have a compass, but you had a small bar magnet or a magnetized needle, what could you do to find your directions? Would you like to create a story about that? Try it. It will be fun.

Teacher Information

The floating needle will behave like a compass, with one end always pointing to magnetic north. A story is told that this is the same type of compass used by Christopher Columbus. Since the metal needles of the time were poor, Columbus had a lodestone to remagnetize his floating needle. A bar magnet will affect the floating needle in the same way it did the compass. By moving the bar magnet under the dish, the children can discover that magnetism goes through water.

Integrating

Social studies

Science Process Skills

Observing, inferring, predicting, communicating

What Are the Earth's Magnetic Poles?

(Classroom demonstration and total-group discussion)

Materials Needed

- 13-cm (6-in.) plastic foam ball cut in half
- Bar magnet
- Compasses
- Toothpicks
- Index card
- Iron filings

For Problem Solvers

Using the Internet and other available reference materials, learn all you can about the earth's magnetic field. When did people first know about the earth's magnetic field? Was it useful information to them? In what ways? Do the earth's magnetic poles always stay at the same place, or do they wander around? Are true north and south and magnetic north and south the same? Do other planets have magnetic fields? Does the moon? What about the sun?

Teacher Information

This activity does not provide procedural steps for students. The concept of the earth's behaving as a large magnet with magnetic north and south poles is difficult to teach with simple inquiry/discovery activities alone. You might want to begin with a classroom discussion using one-half of a plastic foam ball (or a grapefruit or large orange) with a bar magnet running through it and a tip sticking out of each end to represent the magnetic north and south poles. Toothpicks nearby could represent the true north and true south poles, the axis (an imaginary line from true north to true south) on which the earth turns. Research has shown that the north and south magnetic poles wander. Their location is constantly monitored by the Geological Survey of Canada, and exact information of current location and rate/direction of movement can be obtained from their Website.

To further reinforce the idea, place an index card over the bar magnet with iron filings sprinkled on it. Scientists believe this pattern represents approximately the lines of force of the earth's magnetic field, with the strongest pull at the poles. With this basic idea in mind, a compass can be brought near the ball to show that it will point to the north and the south ends or poles.

See the Internet and the encyclopedia for a discussion of northern and southern lights (aurora borealis and aurora australis). Scientists believe the stronger pull at the earth's poles attracts electrons given off from the sun. As they enter the earth's atmosphere, they produce these unusual lights.

The earth's magnetic poles are offset somewhat from the true north and south poles. The difference between true north and magnetic north is called the angle of declination, and it varies according to your geographic location on the earth.

Integrating
Reading, language arts, social studies

Science Process Skills
Observing, inferring, predicting, communicating, researching

How Do Materials Become Magnetized?

Materials Needed

- Mini-bottles, the blanks that are made into two-liter bottles two-thirds full of iron filings (a toothbrush container or test tube will also work)
- Strong magnets
- Compasses

Procedure

1. Place the tube with the iron filings flat on a table.
2. Move the compass along the side of the tube. Observe the needle.
3. Rub the tube containing iron filings thirty times in the same direction across one pole of the strong magnet. Now move the compass along the side of the tube again. Observe the needle.
4. What happened?
5. Shake the tube several times. Move the compass along the side of the tube.
6. What do you think is happening?

For Problem Solvers

Find a screwdriver and try to pick up paper clips with it. If it will pick up one or more paper clips, the screwdriver is already a magnet. If it won't, stroke it the length of the shaft several times in the same direction with a good magnet. Then try to pick up some paper clips with the screwdriver.

What other tools or items do you think you can make into magnets? Wrenches? Scissors? Get permission before you try any of these things, unless they belong to you.

Magnetism

Teacher Information

The "mini-bottle" mentioned in the list of materials looks like a large test-tube with a threaded cap. The 2-liter bottle starts out this way, then with heat and pressure pops out to become the commonly known bottle. Educational Innovations calls them "pop bottle preforms," Learning Resources calls them "plastic test tubes with caps," and Steve Spangler calls them "baby soda bottles."

This activity illustrates in concrete form one of the theories scientists use to explain what happens when an object becomes magnetized. In materials that can be magnetized are groups or domains of atoms that have north and south poles, but they are arranged randomly. (See Figure 6.14–1.)

Figure 6.14–1. Domains Shown in Random Order

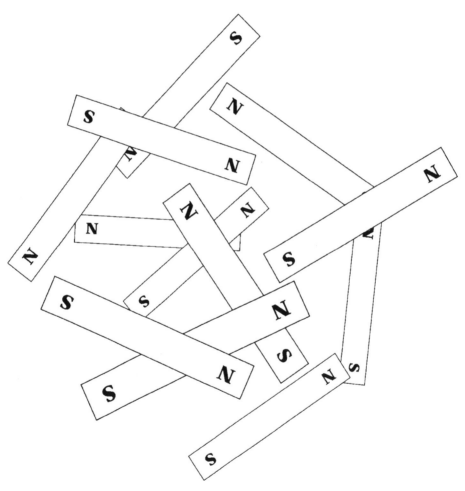

Figure 6.14–2. Domains Lined Up Uniformly

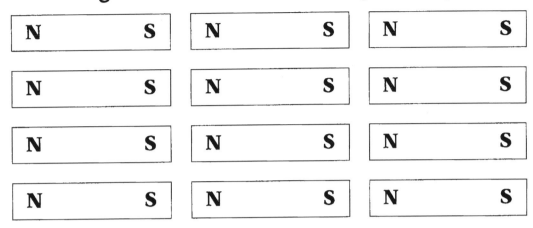

The iron filings in the plastic tube represent these domains. The compass will show that the tube does not have poles. When the domains of atoms come into the presence of a strong magnetic field, they line up, following the lines of force of the magnetic field. (See Figure 6.14–2.)

When the plastic tube is rubbed with a strong magnet, the iron filings line up inside, and the compass test will show the tube has poles. Shaking the tube several times mixes the filings up, and the compass test will show the tube no longer has poles.

The mixing up of the domains is one way of explaining how magnets may lose their magnetism. When an object is heated, the molecules move more rapidly and bounce against one another. Dropping or striking a magnet may jar the domains out of alignment. Magnets stored with like poles together seem to gradually shift the domains out of position.

As the domain theory is presented to children with the plastic tube model, it is important to remember that it is a simplified version. More complex concepts consistent with this model are studied in later years.

Integrating
Math, language arts

Science Process Skills
Observing, inferring, predicting, communicating

How Can You Find the Poles of a Lodestone?

Materials Needed

- Lodestones
- Compasses

Procedure

1. Bring the compass near different parts of the lodestone.
2. Observe the needle.
3. What happened?
4. Discuss your observations and ideas with your group.

For Problem Solvers

Do this project as a Science Investigation. Obtain a blank copy of the "Science Investigation Journaling Notes" from your teacher. Write your name, the date, and your question at the top. Plan your investigation through item 5 (Procedure) and have it approved by your teacher. Complete the Journaling Notes as you perform your investigation. Share your project with your group, and submit your Journaling Notes to your teacher if requested.

Try to identify all the poles that you can on your lodestone. Draw two outlines of the lodestone on paper and label one outline "Front" and the other "Back." Every place that one of the ends of the compass needle is attracted to is a pole. Label each one as S or N. Do the poles seem to be always at the points of the rock, or in the hollows, or both? Do they seem to be always in pairs? How many pairs of poles did you find?

Teacher Information

The compass will indicate that lodestones do have poles. Some lodestones may have several poles, but they should have equal numbers of north and south poles. Some scientists believe that lodestones are magnetized and aligned to the earth's magnetic poles as the iron ore from which they are formed cools.

Since ancient deposits of lodestone have poles pointing in directions different from the present north and south, it is suggested that our present north and south poles may have switched several times over the years.

The different directions in which the poles of ancient deposits point are also studied as indicators of the earth's movement and the theory of continental drift.

Integrating

Language arts

Science Process Skills

Observing, inferring, predicting, communicating

ACTIVITY 6.16

How Can You See a Magnetic Field?

Materials Needed

- Two thin books per small group
- One shaker of iron filings per small group
- Bar magnets
- Cards

Procedure

1. Place the bar magnet between the two books.
2. Cover the magnet with the card, as shown in Figure 6.16–1.
3. Sprinkle iron filings on the card.
4. Tap the card gently several times.
5. What happened?
6. Discuss your observations and ideas with your group.

Figure 6.16–1. Books, Card, Magnet, and Iron Filings

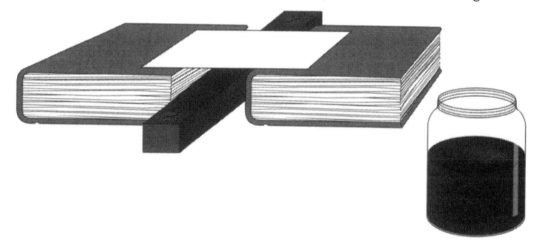

Hands-On Physical Science Activities

For Problem Solvers

Let's explore with magnetic fields. Put a second bar magnet end-to-end with the first one, with unlike poles together, and center them under the card. Sprinkle iron filings on the card and see what the magnetic field looks like now. Next, turn one of the bar magnets around, so the like poles are together, and again sprinkle the card with iron filings. Find other sizes and shapes of magnets and make pictures of their magnetic poles with the iron filings.

Teacher Information

You can demonstrate to the entire class at once by putting the magnets on an overhead projector, using a sheet of glass instead of a card, and shining the image on the screen.

The iron filings will be aligned with the magnetic field of the bar magnet, making it visible. Note that the filings in the middle will point to the poles, while the greater number of filings will cluster at the poles. Iron filings may be purchased from science-supply houses. They also occur naturally in many types of sand and may be collected by running a strong magnet through a sand pile. For easy cleanup, place the magnet in a plastic bag before dragging it through the sand. (If you turn the bag inside out before placing it over the magnet, you can then roll it off the magnet, turning it right-side-out, and the iron filings you collected from the sand are already bagged!)

Your problem solvers will experiment with magnets of different types and will compare the pattern of filings created by their unique magnetic fields.

Integrating

Language arts

Science Process Skills

Observing, inferring, predicting, communicating

How Can You Preserve a Magnetic Field?

Materials Needed

- Two thin books per small group
- Shaker of iron filings per small group
- Bar magnets
- Cards
- Spray paint
- Newsprint or other protective covers

Procedure

1. Do this activity outdoors. You need plenty of ventilation.
2. Place the bar magnet between the two books.
3. Cover the books and magnet with newspaper.
4. Cover the magnet with the card, as shown in Figure 6.17–1.
5. What do you think the pattern of the magnetic field will be like for the magnet(s) you are using? Make your prediction before you sprinkle any iron filings.
6. Sprinkle iron filings on the card.
7. Tap the card gently several times.
8. When the magnetic field is formed, spray the card lightly with paint.
9. Allow the paint time to dry, then brush off the iron filings into a container; they will still work fine as colorful iron filings.
10. Display your picture on a wall, along with those that are made by other students.

For Problem Solvers

Experiment with different magnets and combinations of magnets to create interesting patterns of magnetic lines of force, and paint them so the patterns can be preserved and displayed.

If you can have some spray adhesive, ask your teacher to help you repeat the same activity but use adhesive instead of paint. This time, don't brush off the iron filings when the design has dried. Don't move the card or the magnet until the adhesive is completely dry, which will likely take about thirty minutes. Be sure to do this activity outdoors.

Figure 6.17-1. Newsprint, Card, and Magnet

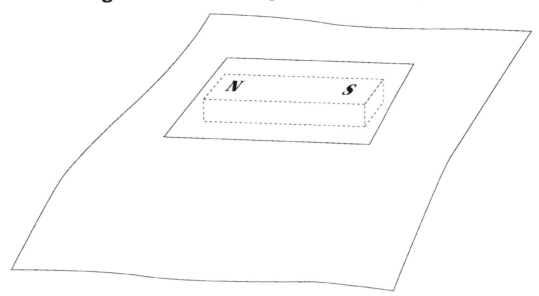

You could also use light-sensitive paper in place of the card to make a permanent record. Even purple construction paper will fade if left by a window in the sunlight and will make a permanent print of the pattern of filings. Where the paper is protected from the sun's rays by the filings, the paper will retain its color, leaving a visible pattern of the filings.

Teacher Information

This is a great way to permanently preserve patterns of magnetic fields for viewing and discussion. Some very interesting patterns will evolve as students become involved. You will get face patterns and many other creative designs as students experiment with a variety of magnets, arranging each one differently in making their patterns.

If spray adhesive is available, help your problem solvers to try at least one of the patterns using adhesive instead of paint. They will be excited to see the iron filings preserved as a part of the pattern. Be sure to let the adhesive dry completely before moving the card or the magnets, and the iron filings that are near the poles will retain their standing and slanting positions. These patterns will need gentle handling to avoid knocking the iron filings off.

Integrating

Art

Science Process Skills

Observing, inferring, classifying, predicting, communicating, identifying and controlling variables

Can You Solve This Magnetism Word Search?

Try to find the following Magnetism terms in the grid below. They could appear in horizontal (left to right), vertical (up or down), or diagonal (upward or downward) position.

magnetize   magnetism   magnetic

lodestone   nonmagnetic   iron

nickel   cobalt   steel

north pole   south pole

```
A  T  U  Y  T  R  E  E  W  Q  C  Z
S  L  I  M  N  B  M  V  C  I  X  N
D  A  O  L  A  K  A  J  T  U  I  O
F  B  P  Y  U  G  G  E  I  P  O  R
N  O  N  M  A  G  N  E  T  I  C  T
G  C  I  T  R  G  E  E  S  W  Q  H
H  X  C  H  A  G  T  F  T  X  Z  P
J  C  K  M  P  K  I  U  E  I  A  O
K  V  E  D  L  K  S  Y  E  R  Z  L
L  B  L  F  H  M  M  T  L  O  M  E
M  N  L  O  D  E  S  T  O  N  E  N
S  O  U  T  H  P  O  L  E  O  L  K
```

Can You Create a New Magnetism Word Search of Your Own?

Write your Magnetism words in the grid below. Arrange them in the grid so they appear in horizontal (left to right), vertical (up or down), or diagonal (upward or downward) position. Then fill in the blank boxes with other letters. Trade your Word Search with someone else who has created one of his or her own, and see whether you can solve the new puzzle.

_____   _____   _____

_____   _____   _____

_____   _____   _____

_____   _____   _____

_____   _____   _____

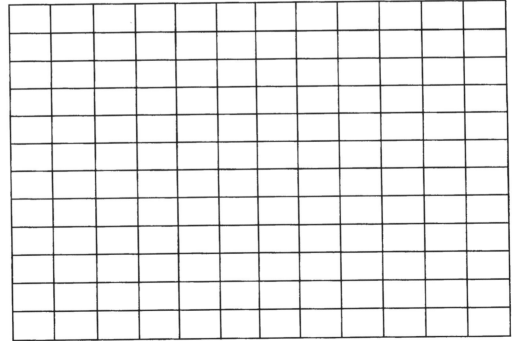

Magnetism

Answer Key for Magnetism Word Search

```
A  T  U  Y  T  R  E  E  W  Q  C  Z
S  L  I  M  N  B  M  V  C  I  X  N
D  A  O  L  A  K  A  J  T  U  I  O
F  B  P  Y  U  G  G  E  I  P  O  R
N  O  N  M  A  G  N  E  T  I  C  T
G  C  I  T  R  G  E  E  S  W  Q  H
H  X  C  H  A  G  T  F  T  X  Z  P
J  C  K  M  P  K  I  U  E  I  A  O
K  V  E  D  L  K  S  Y  E  R  Z  L
L  B  L  F  H  M  M  T  L  O  M  E
M  N  L  O  D  E  S  T  O  N  E  N
S  O  U  T  H  P  O  L  E  O  L  K
```

Do You Recall?

Section Six: Magnetism

1. What materials are magnetic (attracted by a magnet)?

2. What materials can magnetism pass through?

3. Can you tell how strong a magnet is by its size or shape?

4. What part of a magnet has the strongest pull?

5. How do magnets respond to each other?

6. What is a compass needle?

7. Are magnetic lines of force imaginary?

Magnetism

Answer Key for Do You Recall?

Section Six: Magnetism

Answer	Related Activities
1. Iron, steel, nickel, and cobalt	6.5
2. Any nonmagnetic materials	6.6
3. No	6.7
4. Magnets are strongest at the poles.	6.8
5. Like poles repel; unlike poles attract.	6.9
6. A magnetized needle that can move freely	6.10–6.13
7. No. They are invisible, but they are real.	6.16

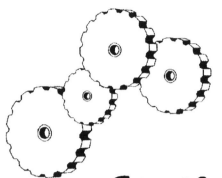

Static Electricity

To the Teacher

The topic of static electricity has at least two things in common with magnetism: (1) the attraction and repulsion of one object for another and (2) the attraction of all age groups to the topic. There is something, perhaps the element of mystery, that intrigues both young and old. Although our information about static electricity is still based largely on theory (another common thread with magnetism), much is known about its behavior. Through observation and experimentation, logical explanations of the phenomenon have developed.

This section is suitable for virtually every grade level, and most activities can be easily adapted for use where needed. The first activity is intended as a teacher demonstration and works well as an introduction to the unit. If used this way, it will raise questions, not provide answers, which is exactly what it is intended to do. A better time to discuss concepts is after students have experimented, discovered, and formed general ideas about the way electrically charged objects seem to behave. Static electricity works best on a cool, clear day (moisture in the air tends to drain off the charge), but success with these activities is very high almost any time except, perhaps, on a hot, muggy day.

Static electricity is the object of curiosity, annoyance, and humor. During a severe electrical storm, the emotion associated with it can be fear, in recognition of the all-too-real danger to life and property. The lightning bolt itself is current electricity because it is moving, but it results from the buildup of an electrostatic charge.

Many of the activities in this section involve objects receiving an electrostatic charge by induction. When two materials rub together, such as wool cloth and a balloon, electrons are transferred from one material to the other—from cloth to balloon, for instance. However, a neutral object can receive an induced charge simply by being near a charged object. If a neutral object is approached by a negatively charged object, the neutral object becomes positively charged. If the neutral object is approached by a positively charged object, the neutral object becomes negatively charged. The induced charge is always opposite that of the charged object nearby. Thus, charged objects are attracted to neutral objects. A charged balloon will cling to a wall, a person, and so forth.

As you select materials to use with static electricity activities, avoid those that have been treated with anti-static chemicals. Such chemicals are often used in dishwashers (anti-spot substances), in clothes dryers (anti-cling products or fabric softeners), or on carpets and furniture (fabric protector). Sometimes these treatments can be washed out, and the materials can then be used for static electricity activities.

Although wool cloth is often suggested for use in static electricity activities because it gives up electrons readily to other materials rubbed by it, cotton cloth, flannel, and many other fabrics work well. Be sure the fabric is clean and is not treated with a fabric softener.

The following activities are designed as discovery activities that students can usually perform quite independently. You are encouraged to provide students (usually in small groups) with the materials listed and a copy of the activity from the beginning through the "Procedure." The section titled "Teacher Information" is not intended for student use, but rather to assist you with discussion following the hands-on activity, as students share their observations. Discussion of conceptual information prior to completing the hands-on activity can interfere with the discovery process.

Regarding the Early Grades

With verbal instructions and slight modifications, many of these activities can be used with kindergarten, first-grade, and second-grade students. In some activities, steps that involve procedures that go beyond the level of the child can simply be omitted and yet offer the child an experience that plants the seed for a concept that will germinate and grow later on.

Hands-On Physical Science Activities

Teachers of the early grades will probably choose to bypass many of the "For Problem Solvers" sections. That's okay. These sections are provided for those who are especially motivated and want to go beyond the investigation provided by the activity outlined. Use the outlined activities and enjoy worthwhile learning experiences together with your young students. Also consider, however, that many of the "For Problem Solvers" sections can be used appropriately with young children as group activities or as demonstrations, still giving students the advantage of an exposure to the experience and laying groundwork for connections that will be made later on.

Correlation with National Standards

The following elements of the National Standards are reflected in the activities of this section.

K–4   Content Standard A: Science as Inquiry

As a result of activities in grades K–4, all students should develop:

1. Abilities necessary to do scientific inquiry
2. Understanding about scientific inquiry

K–4   Content Standard B: Physical Science

As a result of activities in grades K–4, all students should develop understanding of

1. Properties of objects and materials
2. Position and motion of objects
3. Light, heat, electricity, and magnetism

5–8   Content Standard A: Science as Inquiry

As a result of activities in grades 5–8, all students should develop:

1. Abilities necessary to do scientific inquiry
2. Understanding about scientific inquiry

5–8   Content Standard B: Physical Science

As a result of activities in grades 5–8, all students should develop understanding of

1. Properties and changes of properties in matter
2. Motions and forces
3. Transfer of energy

What Is the Kissing Balloon?

(Teacher demonstration)

Materials Needed

- Balloon
- Wool cloth
- String
- Permanent marker
- Words to the story (given below)

Procedure (for teacher)

1. Inflate the balloon.

2. Draw a face on the balloon, using the marker.

3. With the string, suspend the balloon from the ceiling. Adjust the height so the balloon is at the level of your head when you're standing on the floor. Do this while students are out of the room.

4. With students still out of the room, rub the "nose" of the balloon with the wool cloth. If properly charged, the balloon will now face you any time you are reasonably near. If you walk around the balloon, it will follow.

5. You are now ready for your students to return to the classroom. They should take their seats without going near the balloon.

6. Without revealing scientific principles involved, tell the following story as an introduction to your study of static electricity. Following the study, the Kissing Balloon could be used again for review and/or evaluation. A suggestion for such an evaluation follows the story.

"Students, I'd like you to meet a friend of mine. His name is George. George, meet the smartest [fourth] grade in the state."
At this point you are standing at least 2 m (2 yds.) from George.

"There are a couple of things I think you should know about George, class. First, he's nearsighted, and second, he has an awful crush on me. He likes me so much that he just can't keep his eyes off me—when he can see me, that is." (Walk over closer to George and he will turn to face you.)

"You see? He just stares at me, and he keeps staring at me as long as I'm close enough for him to see me." (Walk away.) "When I walk away, he just looks all over, trying to find me again." (Move closer again.) "You'll be able to tell when I'm within his range of vision because he'll look right at me." (Walk around George.) "See, I told you he likes me. Now, if you'll promise not to tell anyone, I'll let George kiss me—just once, on the cheek." (Lean over, where your cheek is very near George.) "Aw, that's so nice.

"Now, for the next few days, we're going to do a few things that should give you some clues about George. We'll visit with him again on another day."

The story can be adjusted to let it be one of the students that George (or Sue, or whatever you want to call ol' Bubblehead) is attracted to. Fun will be had by all, and your students should be more highly motivated for their study of static electricity as a result of meeting George.

One idea for an evaluation following learning activities on static electricity is to repeat the kissing balloon activity, put the following terms on the chalkboard, and ask students to explain why George behaved the way he did. Instruct them to use some of the terms from the chalkboard in their explanation:

- Static electricity
- Attraction
- Repulsion
- Transfer of electrons
- Induction

Note: One of the properties of static electricity is that a charge can be held in a given location. When you rub George's nose with the wool cloth, the charge does not spread throughout the balloon, but remains localized. Otherwise, this activity would not be possible.

Integrating

Language arts

Science Process Skills

Observing, inferring, predicting, communicating, formulating hypotheses

How Does Rubbing with Wool Affect Plastic Strips?

Materials Needed

- Two plastic strips per small group
- Wool cloth

Procedure

1. Hold two plastic strips at one end and let them hang down, face-to-face. What did they do?

2. Place the two plastic strips on the table and rub them with wool, stroking in only one direction. Do not let the plastic strips touch each other.

3. Carefully remove the strips, touching them only at one end. Do not let them touch each other as you pick them up.

4. Now place them together face-to-face, and let them hang down again.

5. What happened? Why do you think they reacted this way?

For Problem Solvers

Do this project as a Science Investigation. Obtain a blank copy of the "Science Investigation Journaling Notes" from your teacher. Write your name, the date, and your question at the top. Plan your investigation through item 5 (Procedure) and have it approved by your teacher. Complete the Journaling Notes as you perform your investigation. Share your project with your group, and submit your Journaling Notes to your teacher if requested.

The variables in this activity are what is being rubbed, how you are rubbing, and what you are rubbing with. Experiment with the variables. Does it have to be these particular strips of plastic? Does it have to be plastic? Do you have to rub with wool cloth? Might cotton cloth work? See what you can learn through your investigation, and share your information with the class.

Teacher Information

To begin with, the plastic strips should both be neutral, and therefore neither attract nor repel each other. When rubbed with wool, which gives up electrons readily, the plastic strips take on a negative electrostatic charge; they gain an excess of electrons. Since the plastic strips are now charged alike, they will repel each other when they are held up.

Integrating

Math, language arts

Science Process Skills

Observing, inferring, classifying, predicting, communicating, formulating hypotheses, identifying and controlling variables, experimenting

What Changes the Way Balloons React to Each Other?

(Take home and do with family and friends.)

Materials Needed

- Two balloons of the same size
- Two pieces of string 60 cm (2 ft.) long
- Wool cloth

Procedure

1. Inflate the two balloons and tie the ends with string.
2. Stand up and hold the two balloons by their strings, away from you, so they hang about an inch or two apart. How do they respond to each other?
3. Rub one balloon with the cloth and repeat step 2. What happened?
4. Rub the other balloon with the cloth and repeat step 2. What happened this time?
5. Explain. Discuss your ideas with your group.

Teacher Information

At step 2, the balloons are neutral and do not react to each other. At step 3, the rubbed balloon has a negative charge and it induces a positive charge in the other balloon. Thus, the balloons attract each other. When the two balloons touch, electrons are transferred from the negatively charged balloon, giving the other one a negative charge, and the balloons repel each other. When both balloons have been rubbed with wool, at step 4, they have like charges (negative) and will repel one another.

Integrating

Math, language arts

Science Process Skills

Observing, inferring, predicting, communicating, comparing and contrasting, formulating hypotheses, identifying and controlling variables, experimenting

What Will a Comb Do to Styrofoam?

(Take home and do with family and friends.)

Materials Needed

- Comb (clean)
- Wool cloth
- Styrofoam® (small piece, as from packing material)
- Thread, about 30 cm (1 ft.) long

Procedure

1. Tie one end of the thread to the Styrofoam and suspend it in the air by the other end of the thread.
2. Rub the solid back of the comb vigorously with the wool cloth.
3. Bring the rubbed part of the comb near the Styrofoam. Hold the comb steady and observe until you see something happen. Observe very carefully, holding the comb steady.
4. In class, explain what you observe and your ideas about why you think it occurred.

Teacher Information

The comb, being rubbed with wool, takes on a negative charge. When it is held near the piece of Styrofoam, the Styrofoam becomes charged positively by induction and will be attracted to the comb. Patience is needed at this point.

As the Styrofoam remains in contact with the comb, some of the negative charge (electrons) will drain off the comb and onto the Styrofoam. The Styrofoam now has a negative charge (the same as the comb) and will leap from the comb. Notice that the Styrofoam does not simply let go, but rather appears to be thrown from the comb by some force. The force is the repelling action of like electrostatic charges for each other.

Note: Be sure the comb is clean. An oily comb will not produce the expected results.

Integrating

Math, language arts

Science Process Skills

Observing, inferring, predicting, communicating, formulating hypotheses, identifying and controlling variables, experimenting

How Can You Make Paper Dance Under Glass?

(Teacher-supervised activity)

Materials Needed

- Sheets of glass (or inverted shallow, clear bowls)
- Plastic bags (clean)
- Two books per small group
- Small bits of paper

Procedure

1. Support the glass by placing a book under each end, as shown in Figure 7.5–1.

Figure 7.5–1. Glass Supported by Books

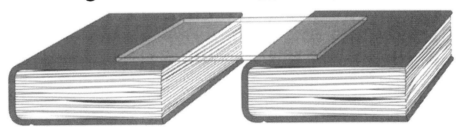

2. Place some small bits of paper under the glass.
3. Rub the top of the glass vigorously with the plastic.
4. Observe for a minute or two.
5. What happened?
6. Why do you suppose it behaves this way?

Teacher Information

Avoid paper that has been treated in any way (for example, to make erasing easy). Tissue paper works well, and plastic foam is an excellent substitute. The glass will give up electrons readily and become positively charged. In turn, it will induce a negative charge in the bits of paper. The paper will jump up to the glass, gradually give up some electrons and fall back to the table where the process of induction is repeated.

Integrating

Math, language arts

Science Process Skills

Observing, inferring, predicting, communicating, formulating hypotheses, identifying and controlling variables, experimenting

ACTIVITY 7.6

Why Does Paper Leap for a Balloon?

(Take home and do with family and friends.)

Materials Needed

- Balloon
- Wool cloth
- Bits of paper (only a few)

Procedure

1. Inflate the balloon and knot the end.
2. Rub the balloon vigorously with the wool cloth.
3. Hold the balloon about 5 cm (2 in.) from the bits of paper and observe for a minute or two. Hold the balloon steady as you observe the bits of paper.
4. What happened? Can you explain why?

For Problem Solvers

By now you know that like charges repel each other and that opposite charges attract each other. For this activity, you charged the balloon by rubbing it with the cloth. The paper was neutral. Why is the paper attracted to the balloon? Do some research about induction, with respect to electrostatic charges. What other activities in this series have materials receiving a charge by induction?

Teacher Information

Tissue paper, or almost any regular writing paper, works very well for the bits of paper, and plastic foam is an excellent substitute.

As the negatively charged balloon approaches the bits of paper, the paper becomes charged positively by induction. The bits of paper are then attracted to the balloon because of opposite charges. When the bits of paper come in contact with the balloon, some of the excess electrons on the surface of the balloon drain off into the bits of paper. The bits of

Static Electricity

paper now have the same charge (negative) as the balloon and repel the balloon. As they drop onto the table top, the excess electrons drain from the bits of paper into the table top, and the process begins again.

Be sure students notice that, as the bits of paper leave the balloon, they don't simply fall, but they are thrown from the balloon by the repelling force of like charges.

Integrating

Math, language arts

Science Process Skills

Observing, inferring, classifying, predicting, communicating, formulating hypotheses, identifying and controlling variables, experimenting

Hands-On Physical Science Activities

What Does Puffed Rice Run Away From?

Materials Needed

- Clear plastic boxes (shallow)
- Styrofoam® (a few small pieces) or vermiculite
- Wool cloth

Procedure

1. Put a few pieces of Styrofoam or vermiculite inside a plastic box.
2. Rub the top of the box with wool.
3. What happened?
4. Bring your finger near the top of the box.
5. What happened?
6. Can you explain your findings? Do you think the Styrofoam is afraid of you?

Teacher Information

This activity involves an induced electrostatic charge beyond that of earlier activities. When the student's finger approaches the negatively charged plastic box, the finger becomes positively charged by induction, just as the Styrofoam inside the box has been. Therefore, the finger and the Styrofoam have like charges. Evidence of the resultant repelling effect is seen as the Styrofoam is "chased" around the box by the finger.

Integrating

Language arts

Science Process Skills

Observing, inferring, classifying, predicting, communicating, formulating hypotheses, identifying and controlling variables, experimenting

How Can You Make Salt and Pepper Dance Together?

(Take home and do with family and friends.)

Materials Needed

- Balloon
- Wool cloth
- Salt and pepper

Procedure

1. Inflate the balloon and tie the end.
2. Sprinkle a small amount of salt and pepper on a sheet of paper or on your desktop.
3. Rub the balloon with the cloth.
4. Bring the balloon within about 2 to 5 cm (1 or 2 in.) of the salt and pepper.
5. Observe for a minute or two.
6. Explain what is happening and why.

Teacher Information

In this activity students will see the salt and pepper do a "dance" because the salt and pepper become charged by induction as the charged balloon approaches. You will need to caution your students to watch very carefully. Otherwise, they will probably not notice that the same grains of salt and pepper are attracted to the balloon, repelled, then attracted again over and over. Notice that they don't simply fall from the balloon but are thrown by the electrostatic force.

Integrating

Math, language arts

Science Process Skills

Observing, inferring, classifying, predicting, communicating, formulating hypotheses, identifying and controlling variables, experimenting

How Can You Make a String Dance?

(Take home and do with family and friends.)

Materials Needed

- Balloon
- String 30 to 45 cm (1 to 1.5 ft.) long
- Wool cloth

Procedure

1. Inflate the balloon and knot the end.
2. Rub the balloon vigorously with the cloth.
3. Lay the string on the table.
4. Bring the balloon near one end of the string, but don't let it touch.
5. What happened?
6. Can you explain why?
7. With practice, you might learn to be a snake charmer!

Teacher Information

The end of the string in this activity is charged by induction and attracted to the balloon. Students enjoy making the end of the string dance.

Integrating

Math, language arts

Science Process Skills

Observing, inferring, predicting, communicating, formulating hypotheses, identifying and controlling variables, experimenting

How Can You Fill a Stocking Without Putting a Leg into It?

(Take home and do with family and friends.)

Materials Needed

- Cool, dry day
- Sheer nylon stocking
- Clean lightweight plastic (such as a produce bag from a grocery store)
- Smooth wall, chalkboard, or corkboard

Procedure

1. Holding it by the top, place the nylon stocking against the wall.
2. Use the plastic to rub and smooth the stocking against the wall. It is best to rub with long strokes in one direction from top to toe. Do this about twenty times.
3. Release your hold on the top of the stocking. What happened?
4. Keeping an arm's length away from the stocking, grasp at the top and slowly pull it away from the wall. Be sure nothing comes near it.
5. Still holding it at arm's length, do you observe any difference in the stocking?
6. Slowly bring the stocking toward you. What happened? Can you explain why this happened?

Teacher Information

This activity demonstrates the attraction and repulsion of like and unlike static charges. Be sure the nylons students use have been washed thoroughly in clear water so any anti-cling treatment has been removed.

At first the nylon will hang limply against the wall. As the plastic is rubbed on the nylon it will remove electrons from the stocking, giving the stocking a positive charge. By the time the stocking has been stroked twenty times, it should smooth out and cling to the wall without support.

When the stocking is pulled away from the wall and held at arm's length, it will fill out in all directions as if an invisible leg were inside. This is because the entire stocking has a positive charge and like charges repel or push away from each other.

Integrating

Math, language arts

Science Process Skills

Observing, inferring, classifying, predicting, communicating, formulating hypotheses, identifying and controlling variables, experimenting

How Can You Bend Water?

(Take home and do with family and friends.)

Materials Needed

- Sink
- Comb
- Wool cloth (or cotton)

Procedure

1. Turn on the water in the sink, just enough to get a very thin but steady stream.
2. Rub the comb vigorously with the cloth.
3. Bring the comb near the stream of water, being sure the comb does not touch the water.
4. What happened?
5. What do you think might have caused this? Discuss your ideas with your group when you are back in school.

Teacher Information

As has been the case with other activities, rubbing the comb with the cloth results in the comb taking on extra electrons and assuming a negative charge. Water molecules are polar, due to the way the two hydrogen atoms combine with an oxygen atom. The result is that a water molecule has a positive charge on one end and a negative charge on the other end. The negatively charged comb attracts the positive end of the water molecule, and the thin stream of water bends noticeably toward the comb.

Integrating

Language arts

Science Process Skills

Observing, inferring, communicating, formulating hypotheses

How Can You Make a Spark with Your Finger?

(Take home and do with family and friends.)

Materials Needed

- Darkened room with carpeted floor

Procedure

1. Turn the lights off and make the room as dark as possible.
2. Shuffle your feet across the carpet for a few steps.
3. Touch a doorknob or some other metal object. Watch carefully at your fingertip as you touch the doorknob.
4. What did you see? If you didn't see anything, try it again.
5. Discuss your observations and ideas with your group.

For Problem Solvers

What is the spark? Do some research on lightning and find out how the spark from your finger might be related to lightning. Is lightning static electricity? Don't jump to conclusions. Study it out.

Find out about lightning safety. What things are important to do during a lightning storm, and what things should you avoid doing?

Teacher Information

This activity is not dangerous, but it does tend to invite horseplay, so if a group of students is involved, close supervision might be needed. The objective of the activity is to see the spark that jumps between the finger and the doorknob after an electrostatic charge is built up from shuffling the feet across the carpet. Students usually discover that a spark can also jump between their fingers and another person's ear or nose. The spark will sometimes be felt and heard. If the room can be darkened, the spark can also be seen. If some students are frightened by the spark, they should not be required to participate.

The spark might be thought of as a miniature lightning bolt. Lightning is a huge spark of electric current that results from a buildup of static electricity in the atmosphere. When moisture in the air and other conditions are just right, in balance with the electrostatic charge, the charge will drain off due to the natural tendency to create electrically neutral conditions.

Integrating

Language arts

Science Process Skills

Observing, inferring, predicting, communicating, formulating hypotheses, identifying and controlling variables, experimenting

Hands-On Physical Science Activities

How Can You Make an Electroscope?

Materials Needed

- Bottles with cork stopper
- Copper wire about 20 cm (8 in.) long
- Lightweight aluminum foil
- Nails
- Rubber combs
- Wool cloth
- Scissors
- Rulers

Procedure

1. Force a nail through the cork stopper to make a hole for the wire.
2. Remove the insulation from both ends of the wire.
3. Insert the copper wire through the cork stopper.
4. Bend the lower end of the wire (the end that will go inside the bottle), as illustrated in Figure 7.13–1.
5. Cut a strip of aluminum foil approximately 1/2 cm (1/4 in.) wide and 3 cm (1.25 in.) long.
6. Fold the aluminum foil in half and let it hang over the end of the wire, as illustrated. Be sure all insulation is removed from the wire where the foil rests.
7. Put the stopper on the bottle, being careful not to jar the strip of aluminum foil off the end of the wire.
8. Rub the comb with the wool cloth and bring the comb near the top end of the wire. As you do this, observe the foil strip carefully.
9. What happened? Can you explain why?
10. Try the same thing with other charged objects.

Figure 7.13–1. Assembled Electroscope

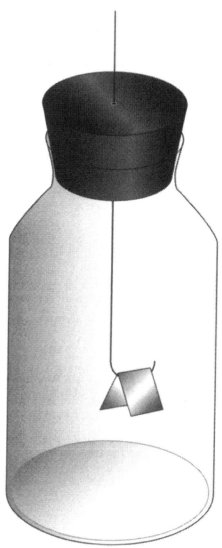

For Problem Solvers

Examine the electroscope and explain why it works the way it does. Share your explanation with at least one of your classmates. When you and your classmates can agree on the way the electroscope works, explain it to your teacher.

Try to make a new design of the electroscope. What are the critical parts? What purpose does the bottle serve? The stopper? The wire? The foil? Can you make something else work in place of any of these? Maybe some of these parts are not even needed at all. Test out your ideas.

Hands-On Physical Science Activities

Teacher Information

The electroscope is an easy-to-make device and should not be difficult to assemble for students who are motivated. They can use it to demonstrate the presence of an electrostatic charge in a comb, a balloon, or other charged object. As the charged object is brought near the upper end of the copper wire, the wire, being a conductor, becomes charged by induction and transfers the charge to the foil. The entire foil strip receives the same charge, and the two ends repel each other.

The very thin foil stripped from a gum wrapper works better for this activity than does heavier aluminum foil used in wrapping food.

Integrating

Math, language arts

Science Process Skills

Observing, inferring, classifying, predicting, communicating, comparing and contrasting, formulating hypotheses, identifying and controlling variables, experimenting

Can You Solve This Static Electricity Word Search?

Try to find the following Static Electricity terms in the grid below. They could appear in horizontal (left to right), vertical (up or down), or diagonal (upward or downward) position.

electricity   static   attract

repel   wool   charge

electron   induction   lightning

conductors   insulators   friction

```
M  N  L  C  H  A  R  G  E  K  J  H
E  I  W  O  Q  A  S  I  D  F  G  T
R  N  R  N  E  T  Y  N  U  I  O  C
H  S  E  D  C  L  J  D  K  L  N  A
F  U  P  U  I  G  E  U  D  O  S  R
E  L  E  C  T  R  I  C  I  T  Y  T
T  A  L  T  A  R  E  T  T  W  A  T
Y  T  U  O  T  I  C  I  O  R  Q  A
N  O  M  R  S  I  L  O  W  O  O  L
B  R  V  S  R  C  X  N  Z  A  S  N
V  S  B  F  H  U  P  O  I  K  L  P
R  F  I  L  I  G  H  T  N  I  N  G
```

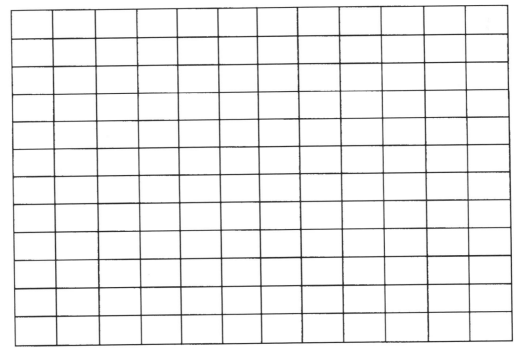

Can You Create a Static Electricity Word Search of Your Own?

Write your Static Electricity words in the grid below. Arrange them in the grid so they appear in horizontal (left to right), vertical (up or down), or diagonal (upward or downward) position. Then fill in the blank boxes with other letters. Trade your Word Search with someone else who has created one of his or her own, and see whether you can solve the new puzzle.

_____   _____   _____

_____   _____   _____

_____   _____   _____

_____   _____   _____

Answer Key for
Static Electricity Word Search

```
M  N  L  C  H  A  R  G  E  K  J  H
E  I  W  O  Q  A  S  I  D  F  G  T
R  N  R  N  E  T  Y  N  U  I  O  C
H  S  E  D  C  L  J  D  K  L  N  A
F  U  P  U  I  G  E  U  D  O  S  R
E  L  E  C  T  R  I  C  I  T  Y  T
T  A  L  T  A  R  E  T  T  W  A  T
Y  T  U  O  T  I  C  I  O  R  Q  A
N  O  M  R  S  I  L  O  W  O  O  L
B  R  V  S  R  C  X  N  Z  A  S  N
V  S  B  F  H  U  P  O  I  K  L  P
R  F  I  L  I  G  H  T  N  I  N  G
```

Do You Recall?

Section Seven: Static Electricity

1. How do two charged objects respond to each other if they have like charges?

2. How do two objects respond to each other if one has a positive charge and the other has a negative charge?

3. Is static electricity a type of magnetism? Explain your answer. (Think about the kinds of objects that are attracted and repelled.)

4. When a charged object approaches a neutral object, what happens?

5. After you slide your feet across an untreated carpet, why do you feel a shock when you reach out to a doorknob?

6. What is an electroscope?

Answer Key for Do You Recall?

Section Seven: Static Electricity

Answer	**Related Activities**
1. They repel each other.	7.2, 7.3, 7.7, 7.10
2. They attract each other.	7.3–7.9
3. No. They have similar behaviors, so they are often confused, but they are different forces.	6.5, 6.9, 7.2–7.9
4. They are attracted to each other.	7.1, 7.3–7.9
5. The excess electrons you pick up from the carpet drain off into the doorknob.	7.12
6. A simple device that detects an electrostatic charge	7.13

Current Electricity

To the Teacher

The study of this topic should follow the studies of magnetism and static electricity, as it requires background information from both. The activities are nongraded, but teachers of young children will need to adapt language and instruction.

The amount of electric current used in these activities (1.5 to 6 volts) is perfectly safe if you follow directions to avoid overheating of circuits. If your flashlight has two batteries, it uses three volts of electricity. *Caution:* Never use household current directly for these activities. In many activities, a high-quality transformer could substitute for batteries. Before using a transformer, be sure you know exactly how much voltage it produces and that it is safe to use. Flashlight cells may be wrapped together securely with electrical tape, end to end, just as they fit in a flashlight. (Two 1.5-volt flashlight cells end-to-end produce 3 volts.)

The majority of these activities suggest using a lantern battery, while a few of them suggest flashlight batteries. The flashlight battery has the advantage of low cost, while the advantages of the lantern battery are

(1) it will last much longer, and (2) it has connectors, making it much easier to connect wires to the batteries (if you get the one with the screw-on connectors rather than the spiral-shaped wire). Battery holders with easy connectors can be purchased or made for flashlight batteries, eliminating that problem.

Be sure that your bulbs are compatible with the voltage of the batteries being used. A bulb designed for 1.5 volts will burn out if you attach it to a 6-volt lantern battery, or even if you connect it to 3 volts, as with two flashlight cells end-to-end.

Technically, a battery consists of two or more cells connected together, and the flashlight battery is not a battery at all, but is more properly called a dry cell. However, single dry cells are so commonly called batteries that to insist on technically proper use of the terms is futile. The popular square 6-volt lantern battery is a true battery, containing four cells of 1.5 volts each (voltage of the four cells is added when they are connected in series within the battery).

Materials such as light sockets, bulbs, and insulated copper wire can be purchased at a hardware or electronics store. An inexpensive wire cutter and stripper (to remove insulation) is a necessary item.

Most activities are designed for individuals or small groups. Setting up learning centers and rotating groups through them can reduce costs by reducing the number of copies of supplies needed for each activity.

Before you begin this section, it would be helpful to preview all of the activities. Some of the same materials are used several times in different ways.

Many schools have magnets that have become weak. You can rejuvenate magnets by wrapping a coil of wire around them and sending an electric current through the wire (see activities on electromagnets). If you recharge bar magnets, be sure the current flows through the wire in the proper direction to produce the correct poles as marked (use a compass to check). A simple device for making bar magnets stronger can be made from a toilet tissue tube with a coil of wire wrapped in one direction around it. Simply insert a bar magnet in the tube and turn on the current for a few seconds. If the poles are reversed, repeat the process but turn the bar magnet around or switch the wires on the battery terminals to send the current in the opposite direction through the coil. Bar magnets can also be given new life by stroking them lengthwise several times across one pole of a powerful magnet. You must stroke in one direction only. If poles are reversed, stroke in the opposite direction or use the other pole of the strong magnet.

Enrichment activities for this unit could focus on new ways to produce electric current. Evaluation should use concrete materials, not just pencil and paper.

The following activities are designed as discovery activities that students can usually perform quite independently. You are encouraged to provide students (usually in small groups) with the materials listed and a copy of the activity from the beginning through the "Procedure." The section titled "Teacher Information" is not intended for student use, but rather to assist you with discussion following the hands-on activity, as students share their observations. Discussion of conceptual information prior to completing the hands-on activity can interfere with the discovery process.

Regarding the Early Grades

With verbal instructions and slight modifications, many of these activities can be used with kindergarten, first-grade, and second-grade students. In some activities, steps that involve procedures that go beyond the level of the child can simply be omitted and yet offer the child an experience that plants the seed for a concept that will germinate and grow later on.

Teachers of the early grades will probably choose to bypass many of the "For Problem Solvers" sections. That's okay. These sections are provided for those who are especially motivated and want to go beyond the investigation provided by the activity outlined. Use the outlined activities, and enjoy worthwhile learning experiences together with your young students. Also consider, however, that many of the "For Problem Solvers" sections can be used appropriately with young children as group activities or as demonstrations, still giving students the advantage of an exposure to the experience and laying groundwork for connections that will be made later on.

Correlation with National Standards

The following elements of the National Standards are reflected in the activities of this section.

K–4   Content Standard A: Science as Inquiry

As a result of activities in grades K–4, all students should develop:

1. Abilities necessary to do scientific inquiry
2. Understanding about scientific inquiry

K-4   Content Standard B: Physical Science

As a result of activities in grades K-4, all students should develop understanding of

1. Properties of objects and materials
2. Position and motion of objects
3. Light, heat, electricity, and magnetism

5-8   Content Standard A: Science as Inquiry

As a result of activities in grades 5-8, all students should develop:

1. Abilities necessary to do scientific inquiry
2. Understanding about scientific inquiry

5-8   Content Standard B: Physical Science

As a result of activities in grades 5-8, all students should develop understanding of

1. Properties and changes of properties in matter
2. Motions and forces
3. Transfer of energy

What Materials Will Conduct Electricity?

Materials Needed

- Circuit as shown in Figure 8.1–1, consisting of a lantern battery, 3 lengths of insulated copper wire of 20 or 22 gauge (insulation removed at the ends), and a small light socket with miniature bulb
- Small objects made of different materials, such as paper clips, nails, wire, wood, rubber bands, glass, plastic, coins, rocks

Procedure

1. Test your circuit by touching the bare wires together. The light should go on.
2. Choose objects from the pile on the table. Touch both bare wires to each object about 2 cm (1 in.) apart and observe the light. If the item conducts electricity, the light will turn on.
3. Explain what is happening the best you can.

For Problem Solvers

Do this project as a Science Investigation. Obtain a blank copy of the "Science Investigation Journaling Notes" from your teacher. Write your name, the date, and your question at the top. Plan your investigation through item 5 (Procedure) and have it approved by your teacher. Complete the Journaling Notes as you perform your investigation. Share your project with your group, and submit your Journaling Notes to your teacher if requested.

Now that you have tested several small items to find out whether or not they conduct electricity, take your circuit and test a lot of other things. For each new material you check, first make a prediction of whether the material will conduct electricity, and then check it out.

Classify the materials you used by making separate lists of materials that conduct electricity, materials that do not conduct electricity, and materials that surprised you. If you found some items that conduct electricity but not very well (the light came on, but it was dim), put them in a separate list.

Teacher Information

This activity introduces the idea of conductors and nonconductors. Non-conductors are also called insulators. Most metals are conductors to some degree, since electricity will move through them. Some materials, such as glass, rubber, and plastic, are insulators, and electricity does not flow through them.

Integrating

Math, language arts

Science Process Skills

Observing, inferring, classifying, predicting, communicating, comparing and contrasting, formulating hypotheses, identifying and controlling variables, experimenting

Figure 8.1–1. Battery, Bulb, and Wires

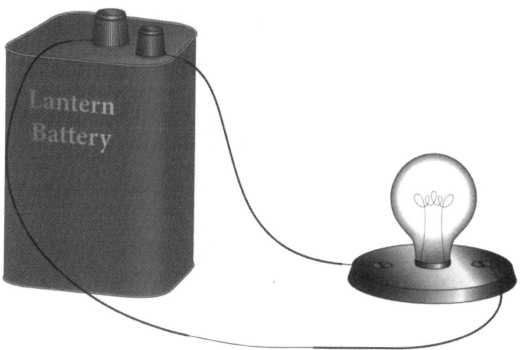

What Is a Circuit?

(Take home and do with family and friends.)

Materials Needed

- One 1.5-volt flashlight battery
- Two 25-cm (10 in.) lengths of single-strand insulated copper wire of 20 or 22 gauge
- Small light socket with flashlight bulb
- Small screwdriver

Procedure

1. In front of you are all the materials needed to make the light turn on. Can you connect them together so the bulb lights up? If you have trouble, read the following hints:

   a. Electricity will flow from the battery (power source) only if it has a complete path from one battery terminal (end) to the other.

   b. Electric current will go through the copper wire, but not through the plastic or rubber insulation (covering) around the wire.

   c. The bulb must be tight in its socket.

   d. The wires must be tightly connected.

2. What happens if the path from one battery terminal to the other is broken? Explain to your partner why you think this happens.

3. When the light is on, this is called a complete circuit because it forms a complete path from the battery to the light and back to the battery.

For Problem Solvers

Remove the bulb from the socket and try to make the bulb light up without the socket. Use only the bulb, two wires, and the flashlight battery.

Make the bulb light by using only one wire.

Now that you know how to make a complete circuit so that the light comes on, try the activity with other people. You only need two wires, a flashlight battery, and a flashlight bulb. Give the materials to your dad, mom, brother, sister, a friend, or your Great Uncle Tom. See whether they can make the light come on. Help them only if you have to, and don't be in too big a hurry to help. Let them struggle for a while. Try it with several different people.

Teacher Information

One and one-half volts of electric current from a flashlight battery is perfectly safe for classroom use. This activity may be varied according to the age of the children. For younger children, you may want to attach the wires to the socket and let them discover how to make it work by screwing the bulb in and touching the bare ends to the battery. Older children should be able to hook up the circuit by following the hints. You may need to help them clean off (strip) insulation from about 2 cm (1 in.) from each end of each wire. An inexpensive wire cutter and stripper obtainable from any hardware store is most helpful.

When completed, the circuit should look like the one shown in Figure 8.2–1.

Figure 8.2–1. Battery and Bulb Connected in a Circuit

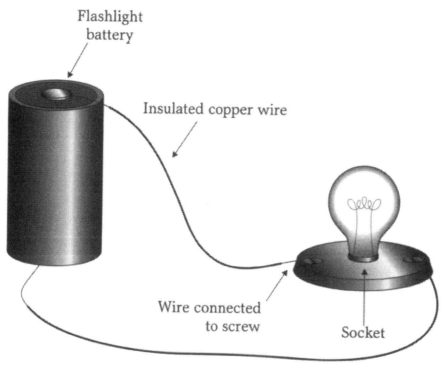

Flashlight battery

Insulated copper wire

Wire connected to screw

Socket

Numbers indicating the gauge (thickness) of wire are in the reverse order of their thickness. For example, 20-gauge wire is heavier than 22-gauge. Either of these will work fine for these activities. You might find a roll of wire labeled as "bell wire" with no gauge identified; it will be fine.

Your problem solvers will have an exciting time challenging others with their newly learned skill. Your students will probably be surprised to learn that few adults can make the bulb light without a bit of struggle. They might even need help, which will do wonders for the self-confidence of the child. The real value will be that the child will learn how to make a complete circuit so well that he or she will probably never forget.

Integrating

Language arts, social studies

Science Process Skills

Observing, inferring, classifying, predicting, communicating, comparing and contrasting, formulating hypotheses, identifying and controlling variables, experimenting, researching

How Can We Model a Complete Circuit?

Materials Needed

- Group of students
- Labels "Battery" and "Light"
- Labels "Negative" and "Positive"
- Paper and pencils

Procedure

1. Assign one person to be the "Battery" and one person to be the "Light."

2. Put the labels on these two people.

3. Put the "Negative" label on the left arm of the "Battery" and the "Positive" label on the right arm of the "Battery." The arms now represent the negative and positive terminals of the battery.

4. Form a circle, with everyone holding hands.

5. Have the "Battery" squeeze the hand of the person on his or her left. That person should, in turn, squeeze with the left hand, and so goes the signal around the circle and back to the positive terminal of the "Battery."

6. This shows a complete circuit, with electricity flowing from the battery, all the way through the circuit, and back to the battery. The "Light" could smile brightly to show that the light turns on when the signal is received. This is called a complete circuit.

7. Have two people—any two—break the hand grip.

8. The "Battery" squeezes with the left hand again, but this time the signal stops where the break is, so it does not get back to the battery.

9. This time the "Light" does not smile, because electricity flows only if it has a complete circuit from the negative terminal of the battery all the way back to the positive terminal of the battery. This time it was an incomplete circuit.

10. Discuss this activity, sharing ideas about what happened and the difference between a complete circuit and an incomplete circuit.

11. Draw a complete circuit and an incomplete circuit, showing the battery, the bulb, and wire.

12. Discuss what you have learned about electricity from this activity.

Teacher Information

It is hoped that this activity will help students begin to understand the difference between a complete circuit and an incomplete circuit and that electricity flows only if the circuit is complete.

In a good battery, chemicals react together in such a way that an abundance of electrons builds up at the negative terminal, with a shortage of electrons resulting at the positive terminal. That's why the terminals are called negative and positive. The natural tendency is toward neutrality, so when a conductor connects the negative terminal to the positive terminal, electrons flow through the conductor, from negative to positive, until the substances inside the battery are electrically neutral. Whether there is a light or other appliance in the circuit is immaterial to whether or not electrons will flow. We simply place an appliance in the circuit to take advantage of the fact that the battery will send electrons from the negative terminal to the positive terminal if the two terminals are connected.

If we remember that the reason electrons flow is to reduce the excess at the negative terminal and make up the shortage at the positive terminal, it's easy to understand why the circuit must be complete in order for electricity to flow. And incidentally, if the battery is still in good condition, the chemical reaction that caused the imbalance of electrons at the two terminals will continue and the battery will partially recover. It might take several weeks, but if it isn't too old it will happen. A standard battery has a zinc case, which slowly deteriorates with time. When the zinc has deteriorated to a certain point, the battery is truly "dead."

The newer long-life batteries use different chemicals and their recovery time will likely be many months, or even years! Rechargeable batteries use still different chemicals, and they do not recover except with a charger.

Integrating

Language arts

Science Process Skills

Observing, inferring, communicating, comparing and contrasting, using space-time relationships, formulating hypotheses

What Is a Short Circuit?

Materials Needed

- Complete circuit used earlier
- One 25-cm (10-in.) length of copper wire, stripped on both ends

Procedure

1. Connect your circuit so there is a complete path and the light turns on.

2. Strip (clean off) the insulation from 1 cm (1/2 in.) in the center of each wire. Does the light still go on?

3. Put the bare ends of the extra piece of wire across the bare sections of your circuit wires, as in Figure 8.4–1. Does the light go on?

4. What do you think has happened? Why?

5. Feel where the bare ends are touching. Do you notice anything?

6. You have made a short circuit. Discuss with your teacher and the other class members what that means.

Figure 8.4–1. Battery and Bulb with Extra Wire

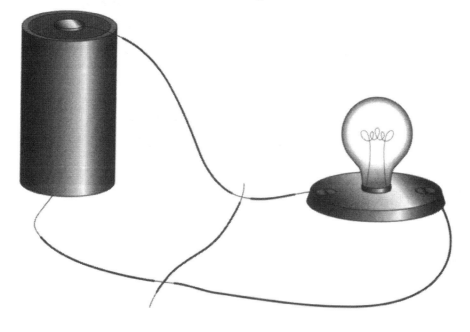

Teacher Information

When the additional wire is placed across the circuit, the light goes out. A general rule in electrical circuits is that electricity will follow the path of least resistance. This means it will follow the shortest, easiest path back to its source (battery). In this example, the path is shorter and avoids a resistor (the light bulb). If you permit current to flow through the short circuit for more than a brief moment, the wire will begin to heat up. This is caused by providing an easy path for electricity to flow. The amount of electricity going through the wire increases beyond its normal capacity.

Caution: Do not leave the circuit connected this way for more than a few seconds. If the battery is strong, the wire will likely become hot enough to burn fingers, although it will still not cause an electrical shock. It will also run the battery down very rapidly.

Integrating

Math, language arts

Science Process Skills

Observing, inferring, predicting, communicating, comparing and contrasting, formulating hypotheses, identifying and controlling variables, experimenting

How Can You Make a Switch?

(Teacher-supervised activity)

Materials Needed

- Battery
- Bulb and base
- Two pieces of insulated wire, about 30 cm (1 ft.) long, with insulation removed from the ends
- One piece of insulated wire, about 45 cm (1.5 ft.) long, with insulation removed from the ends
- Metal strip 10 cm (4 in.) long by 2 cm (1 in.) wide
- Strip of wood (lath) 15 cm (6 in.) long
- Two small wood screws
- Hammer and nail
- Screwdriver

Procedure

1. Use the hammer and nail to punch a hole near one end of the metal strip.
2. Use a screw to attach the metal strip near one end of the piece of wood (if you also punch a small nail hole in the wood, the screw will go in more easily).
3. Install the other screw in the wood so that the unattached end of the metal strip reaches it. Bend the metal strip up about 1 cm (½ in.) so the metal strip is above the screw.
4. Use the long wire to connect the battery to the bulb.
5. Use one of the short wires to connect the bulb to one end of the switch.
6. Use the other short wire to connect the battery to the other end of the switch.
7. When you finish, your circuit should look something like Figure 8.5–1.
8. Can you make the light turn on and off by using the switch?
9. Discuss what the switch does.

Figure 8.5–1. Battery and Bulb with Switch in Circuit

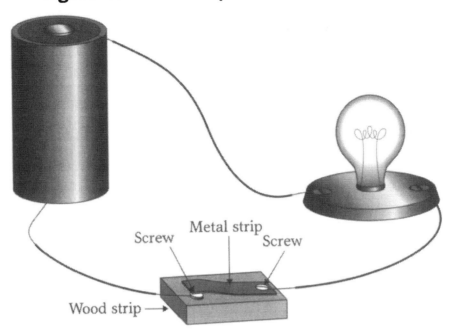

Screw Metal strip Screw
Wood strip →

For Problem Solvers

In this activity you made a switch. An electrical switch is a device that allows you to easily turn electricity on and off. There are many ways you can make a switch. What other ways can you think of to make one? Can you make a switch out of a paper clip? Make several different kinds and test them in your circuit.

Teacher Information

If available, sheet copper should be used because it is easier to cut than sheet metal, and there is less chance of injury from sharp edges and corners. The wires should be under the screw heads when the screws are tightened down.

If you have an electric drill, you might choose to use it instead of the hammer and nail to put the screw hole in the switch base. *Caution:* If you use a drill, be sure to hold the metal strip with pliers to avoid injury to the fingers, as the drill bit will sometimes grab the metal strip and jerk it out of your hand.

This circuit has the following parts:

a. A power source (battery)
b. A path through which the current can flow (wire)
c. An appliance (light) to use current
d. A switch to turn it on and off

Integrating
Math, language arts

Science Process Skills
Observing, inferring, measuring, predicting, communicating, identifying and controlling variables, experimenting

How Can You Make a Series Circuit?

Materials Needed

- Lantern battery with screw-on caps on the terminals
- Circuit from Activity 8.5, including switch
- Two additional small light sockets with flashlight bulbs
- Two more pieces of insulated single-strand copper wire 10 cm (4 in.) long
- Small screwdriver

Procedure

1. For this and some of the following activities, you will use a lantern battery instead of a flashlight battery. It is easier to use and will last longer.

2. In this activity, you will add two lights to your circuit.

3. Test to be certain your circuit will light the bulb.

4. Disconnect one wire from your light socket and connect it to one terminal of another socket with one 10-cm length of wire. Use another 10-cm length of wire to connect the other terminal of your second socket to the third socket. Connect the long wire from the third socket to the battery. When you finish, your circuit should look like Figure 8.6–1.

5. Now close the switch. Do the lights come on? Unscrew one bulb. What happened? Can you explain why?

6. Discuss your ideas with your group.

Figure 8.6–1. Battery, Switch, and Three Bulbs Wired in Series

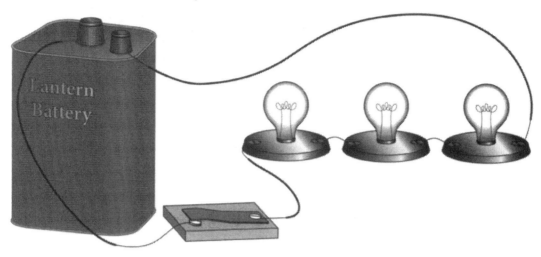

Teacher Information

Before letting the children use the lantern batteries, you may want to remove the screw caps from the terminals and remind the children not to leave the wires connected when they are not in use. Better yet, always use a switch in the circuit, so the switch will take care of this problem by turning the light off when you release the switch. You can then leave the caps on.

Lantern batteries have an advantage over flashlight batteries of being easier to handle and simpler to hook up. There is no danger of electrical shock from these batteries.

Be certain the wires are stripped on both ends. The circuit the students have constructed is called a series circuit because the electric current travels through the wire in one path and the resistors (lights) and switch are all part of that single path. If one portion of a series circuit is missing, the current will not flow, because it must have a complete path from the power source to the power source. If a light bulb is removed or burned out, the path is broken and all the lights go out.

Integrating

Math

Science Process Skills

Observing, inferring, measuring, predicting, communicating, comparing and contrasting, formulating hypotheses, identifying and controlling variables, experimenting

How Can You Make a Parallel Circuit?

Materials Needed

- Series circuit from Activity 8.6
- Screwdriver
- Wire stripper
- Two additional pieces of insulated single-strand copper wire 10 cm (4 in.) long

Procedure

1. Use the additional pieces of wire to change your circuit so it looks like the one in Figure 8.7–1.
2. Close the switch so all bulbs are lit. Unscrew one bulb. What happened? Unscrew two bulbs.
3. Can you explain how this circuit is different from the one in Activity 8.6? Discuss your ideas with your group.

Figure 8.7–1. Battery, Switch, and Three Bulbs in Parallel

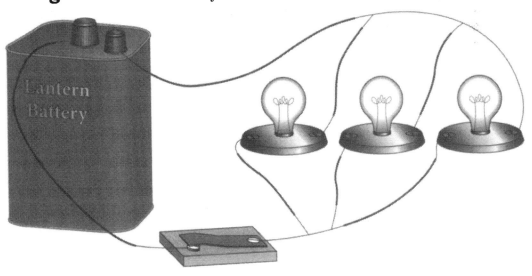

For Problem Solvers

As you have learned about series and parallel circuits, you learned that an important difference is that, if lights are wired in series, they all go out if one goes out. If they are wired in parallel, one light can go out and the rest of them stay on. Sets of holiday lights that are wired in series can be very frustrating to the user. When one bulb burns out, it's very difficult to tell which bulb is bad, because they are all off. With sets of lights that are wired in parallel, it is easy to identify the bulb that needs to be replaced. If you have a choice, which will you buy?

Teacher Information

A parallel circuit provides an independent path for the electric current to travel to each light and back to the power source. The lights will burn brighter because the current does not have to travel through a series of resistors (the other lights). And when one light goes out, the rest of the lights stay on! For many years, the manufacture of holiday lights wired in series has been a great frustration to users of these lights.

Integrating

Reading, language arts, math, social studies

Science Process Skills

Observing, inferring, classifying, predicting, communicating, comparing and contrasting, formulating hypotheses, identifying and controlling variables, experimenting, researching

What Is Resistance?

Materials Needed

- Lantern battery
- Insulated copper wire of 20 or 22 gauge
- Small light socket with miniature bulb
- Writing pencil with about 8 cm (3 in.) of wood removed on one half to expose the graphite core

Procedure

1. Set up a circuit similar to the one shown in Figure 8.8–1.
2. Touch the ends of the bare wire to the graphite ("lead") core of the pencil, as far apart as possible.
3. Slowly slide the bare wires closer together along the graphite and observe the light.
4. Can you use the information you have learned about conductors to explain this?

Figure 8.8–1. Battery, Bulb, and Pencil

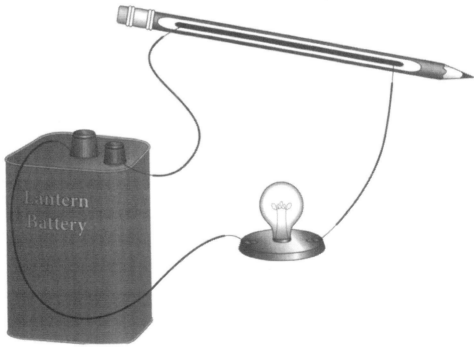

5. Have you seen anything like this principle used in a home or an automobile?

6. Share your ideas with each other.

For Problem Solvers

If you have a switch at home that dims the lights, it works in about the same way as the pencil lead (graphite) you used for this activity. A resistor switch is called a *rheostat.* This device makes it possible to change the amount of current going to an appliance (light, heater, or other item). Make a list of all the things you can think of that use resistor switches. Think about the kitchen. Think about the car. Share your list with others who are doing this activity and with your teacher.

Teacher Information

This is a demonstration of variable resistance. The core of a "lead" pencil isn't lead at all, but graphite. Graphite will conduct electric current, but not nearly as well as copper. Because it resists the flow of current, it is an example of one kind of resistor. As the bare copper wires are moved closer together, the resistance is gradually reduced and the light becomes brighter. This is the principle of the rheostat used to dim automobile dash lights and some lights in homes, theaters, and public buildings.

Integrating

Language arts, social studies

Science Process Skills

Observing, inferring, classifying, predicting, communicating, comparing and contrasting, formulating hypotheses, identifying and controlling variables, experimenting, researching

How Does Electric Current Affect a Compass?

(Teacher-supervised activity)

Materials Needed

- Insulated 22-gauge copper wire, 50 cm (20 in.) long, stripped on each end
- Lantern battery
- Compass

Procedure

1. From your study of magnetism, do you remember that a compass is a magnet suspended so it can turn freely? The magnetic field of the earth causes the compass needle to point north unless another magnet or magnetic material comes near it.
2. Put your compass flat on the table and notice which direction it is pointing.
3. Connect one end of the wire to a terminal of the battery and place the wire across the top of the compass.
4. Touch the other end of the wire to the second terminal of the battery (but don't connect it), as seen in Figure 8.9–1.

Figure 8.9–1. Battery, Wire, and Compass

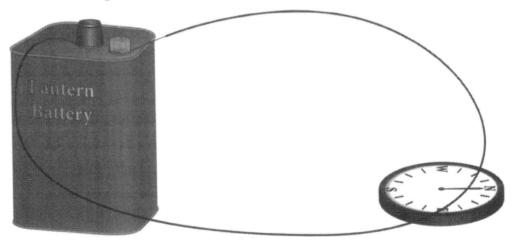

5. Move the wire to different positions on top of the compass and observe the needle as you send current through the wire.

6. What is happening? Discuss this with the class.

Teacher Information

When an electric current flows through a wire, a magnetic field is formed around the wire. This is shown by the reaction of the compass needle when a nearby wire has current running through it.

Caution: When the copper wire is connected, it creates a short circuit. The wire should be connected only momentarily; otherwise, it will get very hot, the battery power will drain rapidly, and fingers can get burned. Many fires are started by short circuits.

Integrating

Math, language arts

Science Process Skills

Observing, inferring, predicting, communicating, formulating hypotheses, identifying and controlling variables, experimenting

What Happens When Electric Current Flows Through a Wire?

Materials Needed
- Circuit used in Activity 8.9
- Iron filings
- 5- by 7-inch card

Procedure
1. Connect one end of the wire to a terminal of the battery.
2. Place the card flat and level over the middle part of the wire.
3. Touch the other end of the wire to the battery terminal and quickly sprinkle iron filings on the card, as shown in Figure 8.10–1.
4. Disconnect one wire from the battery and observe the card. Do not leave the wire connected to the battery for more than a few seconds.
5. Have you seen something like this before? Look carefully at the filings.
6. What conclusions can you make? Discuss this with the class.

Figure 8.10–1. Battery, Wire, Card, and Iron Filings

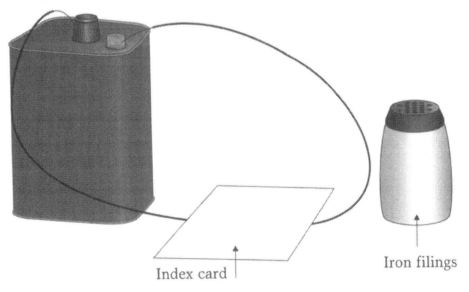

Index card

Iron filings

Teacher Information

When current flows through the wire, the iron filings will line up along the path of the wire because of the magnetic field created around it. Note that iron filings do not line up with the wire, but perpendicular to it instead. This indicates that the lines of force of the magnetic field are perpendicular to the direction of the flow of current.

Integrating

Math, language arts

Science Process Skills

Observing, inferring, predicting, communicating, formulating hypotheses, identifying and controlling variables, experimenting

What Is an Electromagnet?

Materials Needed

- 1 m (1 yd.) of 22-gauge insulated copper wire stripped on both ends
- Lantern battery
- Iron nail, about 10 cm (4 in.) or so
- Paper clips

Procedure

1. Coil the wire ten times around the nail. Bring the nail near some paper clips, as illustrated in Figure 8.11–1. What happened?

Figure 8.11–1. Electromagnet with Nail and Paper Clips

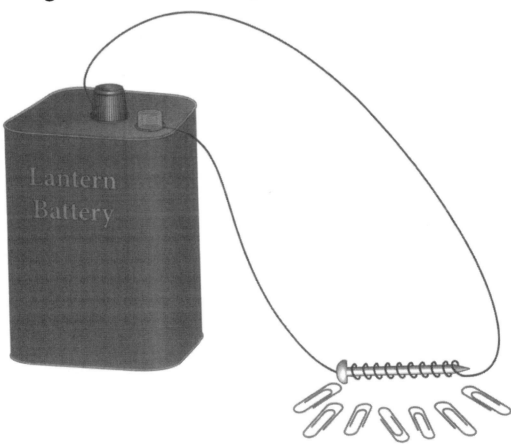

2. Attach one end of the wire to a terminal of the battery. Bring the nail near some paper clips. Touch the other end of the wire to the other terminal of the battery. What happened?

3. Hold the nail above the table and remove the wire from one terminal of the battery.

4. What happened?

5. Explain to each other why you think this happened.

Teacher Information

The nail will not pick up paper clips until electric current flows through the coil of wire. When current flows through the coil of wire, a magnetic field is created around the coil and causes the iron nail to become a temporary magnet and attract the paper clips. When the electric current is cut off, the nail will lose its magnetism and the clips will fall off.

Integrating

Math

Science Process Skills

Observing, inferring, comparing and contrasting, identifying and controlling variables, experimenting

What Is a Way to Change the Strength of an Electromagnet?

Materials Needed

- Same as for Activity 8.11

Procedure

1. As you perform the activities with electromagnets, remember not to leave the electromagnets connected to the battery for more than a few seconds at a time.
2. Use your electromagnet with ten coils to pick up as many paper clips as you can at one time. Record the number of clips it picked up.
3. Wrap ten more coils of wire around the nail, connect it to the battery, and see how many clips you can pick up at one time. Record the number.
4. What do you think would happen if you wrapped ten more coils around your electromagnet? Try it.
5. Discuss your ideas with your group.

Teacher Information

Increasing the number of coils of wire will increase the strength of the magnetic field. This is one way to make an electromagnet stronger. Using a stronger electric current will also increase the force of an electromagnet.

Integrating

Math

Science Process Skills

Observing, inferring, measuring, predicting, communicating, comparing and contrasting, formulating hypotheses, identifying and controlling variables, experimenting

What Is Another Way to Change the Strength of an Electromagnet?

Materials Needed

- Two lantern batteries
- 1 m (1 yd.) of insulated copper wire
- One 10-cm (4-in.) length of insulated copper wire stripped at both ends
- One large nail
- Box of paper clips
- Paper and pencils

Procedure

1. Make a coil of wire by wrapping the 1-meter wire around the nail ten times.

2. Make an electromagnet by connecting the coil of wire to one of the batteries, and see how many clips you can pick up at one time. Record the results. Disconnect one end of the wire from the battery except for the moment that you are using your electromagnet to pick up the paper clips.

3. Observe your lantern battery. Notice it has a terminal (connector) in the center and one near the outside edge. Unless otherwise marked, the center terminal is called positive (+) and the outer terminal is called negative (−).

4. Use the 10-cm wire to connect a second battery in series (connect the wire from the negative terminal of one battery to the positive terminal of the other battery). This increases the voltage to the sum of the voltage of the two batteries.

5. Connect your electromagnet as shown in Figure 8.13–1 and pick up as many clips as you can. Record the number of clips.

Hands-On Physical Science Activities

Figure 8.13-1. Electromagnet with Two Batteries and Paper Clips

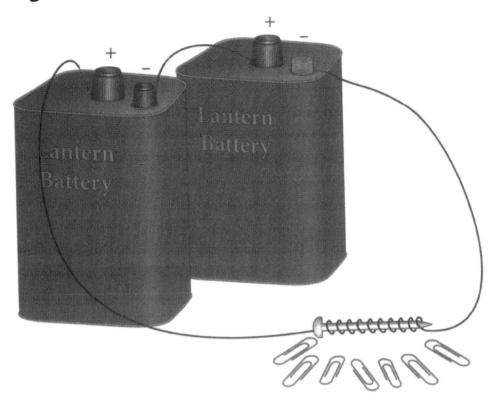

6. What do you think would happen if you used two batteries and wrapped the wire 20 times around the nail? Try it. Be sure to disconnect one end of the wire from the battery as soon as you've picked up the paper clips.

For Problem Solvers

You have now learned two ways to change the strength of an electromagnet. Earlier you learned how to make a switch. Build an electromagnet with a switch in the circuit so that it's easy to turn the magnet on and off. Try different sizes and types of nails and bolts as the core. See whether you can find a brass bolt to try as the core. Find a steel bolt to use instead of iron. Find out what works best and what doesn't work so well.

Use the shaft of a screwdriver as the core of an electromagnet. How does it work differently from the nail or bolt? Does it release all the paper clips when you turn off the power? Oops! The screwdriver is now a permanent magnet!

Do some research about electromagnets. Make a list of some ways they are used.

Teacher Information

Series wiring is commonly used to increase the voltage of dry-cell batteries. Six-volt batteries contain four cells. Each cell produces 1.5 volts, and when they are connected in series, the voltage is added. More about batteries and how they work is developed later in this section. *Warning:* Fingers can get burned if the wire is attached to the battery for more than a few seconds. It will also drain the energy quickly from the battery. Disconnecting one end of the wire from the battery will stop the flow of electricity through the wire.

Integrating

Math

Science Process Skills

Observing, inferring, measuring, predicting, communicating, comparing and contrasting, formulating hypotheses, identifying and controlling variables, experimenting

What Happens When Current Flowing Through a Wire Changes Direction?

Materials Needed

- 50-cm (20 in.) length of insulated copper wire stripped at both ends
- Lantern battery
- Compass

Procedure

1. Place the center of the wire over the compass. Touch the ends of the wire to the terminals of the battery and watch the needle.

2. Now touch the ends of the wire to the opposite terminals of the battery. Since electricity flows from the negative (–) terminal to the positive (+), you have reversed the flow of current through the wire. (See Figure 8.14–1.)

Figure 8.14–1. Compasses Near Wires with Current Flowing in Opposite Directions

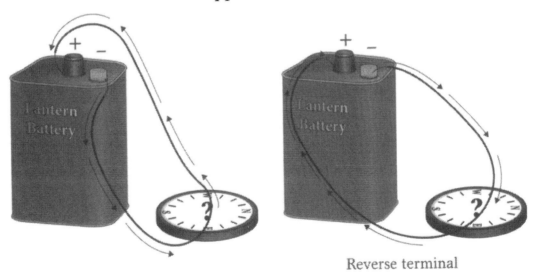

Reverse terminal

3. Switch the ends of the wires several times. What happens to the compass needle?

4. Discuss your observations with your group.

Teacher Information

At this point, you should emphasize that electricity produced by a battery flows in one direction through the wire. It flows from the negative terminal of the battery to the positive terminal when a path is provided. This is called *direct current*, as opposed to house current, which is *alternating current*.

Each time the direction of the flow of current changes in this activity, the compass needle will reverse, indicating that the poles of the electromagnetic field reverse when the direction of the flow of current is reversed.

Integrating

Math, language arts

Science Process Skills

Observing, inferring, communicating, comparing and contrasting, formulating hypotheses, identifying and controlling variables, experimenting

How Does the Direction of the Flow of Current Affect an Electromagnet?

Materials Needed

- Electromagnet used in Activity 8.13
- Lantern battery
- Compass
- Two bar magnets
- String 20 cm (8 in.) long

Procedure

1. *Important:* Throughout this activity, always disconnect one wire from the battery except for the moment that you are using the electromagnet.

2. Place the compass flat on the table. Connect your electromagnet to the battery and bring one point of it near the compass. Observe which end of the needle points to the electromagnet.

3. Switch the ends of the wires leading to the battery terminals. What happened to the compass needle?

4. Suspend one bar magnet from the piece of string. Bring each end of the other bar magnet near the N end of the suspended magnet. What happened? This should help you remember a characteristic of magnets that you learned when you were studying magnetism.

5. Bring your nail and coil of wire near both ends of the suspended bar magnet (don't connect it to the battery yet). What happened?

6. Connect the wires to the battery and bring the pointed end of the nail near the N end of the suspended bar magnet. (See Figure 8.15–1.)

7. Now turn your electromagnet around so the flat end of the nail comes near the N end of the suspended bar magnet. What happened?

Current Electricity

Figure 8.15–1. Setup for Activity 8.15

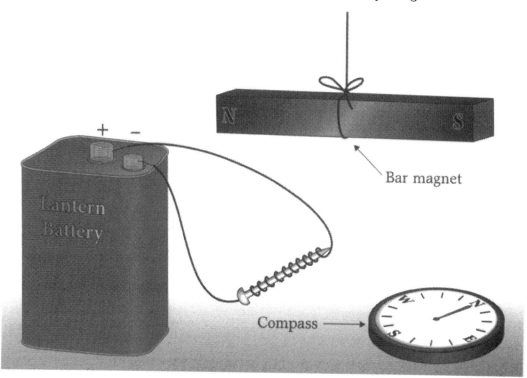

Bar magnet

Compass →

8. Keep the flat end near the suspended bar magnet and reverse the flow of current through your electromagnet by switching the wires on the battery terminals. Do this several times and observe the behavior of the bar magnet (don't move the electromagnet).

9. Describe your observations. What can you say about this characteristic of an electromagnet?

10. Share your ideas with each other.

For Problem Solvers

You have discovered a very important relationship between magnetism and electricity. All electric motors work on two principles that you have learned: first, that a flow of electricity through a wire creates a magnetic field around the wire, and second, changing the direction of the current reverses the poles of an electromagnet.

Now do some research and see how much you can learn about electric motors and how they work on these principles of magnetism and electricity. You might even want to try building a simple electric motor that actually runs.

Teacher Information

Like permanent magnets, electromagnets have poles. When the direction of the flow of current is reversed (in this case by switching wires on the terminals), the poles of the electromagnet reverse. This characteristic of electromagnets gives them an advantage over permanent magnets in some applications. Because of this phenomenon, electric motors are possible. Electric motors are magnetic motors, but without the ability to instantly reverse the poles of magnets, the electric motor would not be possible. To demonstrate this try the following.

Have one student hold the electromagnet near the suspended bar magnet. Have another student touch the wires to the battery. As soon as the poles of the two magnets come close to each other, switch the wires on the terminals of the battery. The pole of the electromagnet, which was attracting a pole of the bar magnet, will reverse and repel the same pole of the bar magnet (remember that opposites attract, likes repel). With practice, the students can make the bar magnet spin around by reversing the poles of the electromagnet. Now all that is needed is a device to switch the direction of current automatically and you have a simple electric motor.

Integrating

Math, language arts

Science Process Skills

Observing, inferring, communicating, comparing and contrasting, formulating hypotheses, identifying and controlling variables, experimenting

How Can You Tell Whether Electric Current Is Flowing Through a Wire?

(Teacher-supervised activity)

Materials Needed

- 10 m (11 yd.) of 24-gauge insulated copper wire stripped at both ends
- Four 10-cm (4-in.) lengths of wire
- One 30-cm (12-in.) length of wire
- Sewing needle (with point broken off)
- One 20-cm (8-in.) strip of wood lath
- Lantern battery
- Thread
- Paper clip
- Bar magnet

Procedure

1. Make a coil with your long piece of wire by winding it around a lantern battery or small fruit jar. Leave about 50 cm (20 in.) of wire at each end.

2. Remove the coil from the battery and use the 10-cm (4-in.) pieces of wire to hold the coil together in four places.

3. Use the 30-cm (12-in.) wire to secure the coil in an upright position on the lath.

4. Magnetize the needle by rubbing it thirty times in the same direction with a bar magnet. It should now pick up a paper clip.

5. Use thread to hang the needle balanced in the middle of the coil. Compare your finished product with Figure 8.16–1.

6. We have learned that electric current flowing through a wire creates a magnetic field around the wire. We found that when the wire is coiled, the magnetic field is stronger. The needle is a magnet. If an electric current flows through the coil of wire, what do you predict will happen to the needle?

474

Figure 8.16–1. Coil of Wire with Suspended Needle

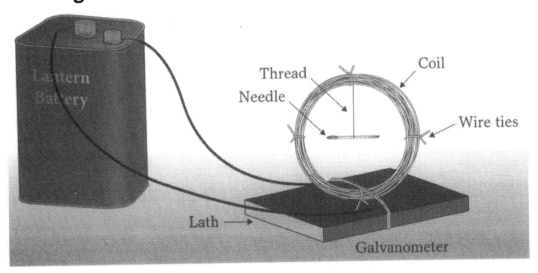

7. Touch the bare ends of your coil of wire to the dry cell. Was your prediction correct?

8. This is a simple galvanometer, used to detect the direction and flow of electric current. Touch the wires to a battery and notice the direction the needle points. Now, switch the wires on the terminals of the battery and notice the direction the needle points. This device can be used to detect the flow of even small amounts of electric current.

9. In what situations do you think such a device as this would be useful? Share your ideas.

For Problem Solvers

In this activity, you made a galvanometer. Design your own galvanometer. Try more windings in the coil. Try fewer windings in the coil. Try placing a directional compass within the coil instead of a suspended needle on a thread. What other variables can you think of to change it and perhaps make it better or easier to use?

Teacher Information

Caution: Even with the point broken off, the needle could injure someone. Careful supervision is required. You might choose to use the compass instead of a magnetized needle, as suggested in "For Problem Solvers."

Ideally, each child should construct a galvanometer. Groups of three or four students working on one galvanometer should be maximum. If a student is able to construct, explain, and predict the behavior of a galvanometer, it will indicate that he or she understands some of the important ideas about magnetism and electricity.

Your problem solvers will substitute a directional compass for the suspended needle.

Very thin insulated wire is best for this activity in order to provide many windings without excessive bulk. The more windings the galvanometer has, the more sensitive it will be to small amounts of current electricity.

Integrating

Math, language arts

Science Process Skills

Observing, inferring, communicating, comparing and contrasting, formulating hypotheses, identifying and controlling variables, experimenting

How Is Electricity Produced by Chemicals?

Materials Needed
- Wide-mouthed glass cup or jar
- 8-cm (3-in.) zinc strip
- 8-cm (3-in.) copper strip
- Salt
- Lemon juice
- Water
- Galvanometer from Activity 8.16

Procedure
1. Pour water into the jar or cup until it is three-fourths full.
2. Dissolve two teaspoons of salt in the water.
3. Tightly connect one end of your galvanometer to the copper strip and put the copper strip in the glass. Bend the zinc strip into a hook on one end and hang it inside the glass, as illustrated in Figure 8.17–1.

Figure 8.17–1. Homemade Cell and Galvanometer

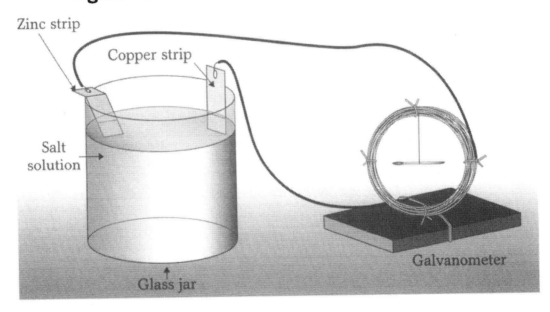

4. Wait a few seconds. Then observe the needle of your galvanometer as you firmly touch the other wire to the zinc strip.

5. What happened? Wait a few seconds, then touch the wire to the zinc again.

6. Discuss your observations with your group.

For Problem Solvers

Try vinegar as the electrolyte instead of salt water, and do the activity again. (Be sure to rinse the jar thoroughly whenever you change the electrolyte.) Try lemon juice. Replace the copper strip with the carbon rod. Try a penny in place of the copper strip and a silver dime instead of the zinc. (You'll have to find a dime that's several years old in order to get silver. If necessary, visit a coin shop.)

Teacher Information

When placed in an acid or base (alkaline) solution, some materials give up or take on electrons readily through a chemical process called ionization. Zinc metal builds up electrons and becomes the negative terminal. When a negative material (zinc) is connected to a positive material (copper or carbon) in the solution, electrons flow along the path from the zinc to the copper or carbon. In this circuit, the galvanometer is in the path and detects the flow of current. You can explain this using the diagram in Figure 8.17–1 by placing "+" beside the copper or carbon rod and "–" beside the zinc strip.

The reason the children are asked to wait a few seconds before touching the wire to the zinc is to allow time to build opposite charges on the copper and zinc strips.

Carbon rods and zinc strips may be obtained by dismantling old-style flashlight batteries. A carbon core runs through the center, and the case is made of zinc. The newer long-life batteries are constructed differently, although the same basic principles apply as they develop an electrical charge. Six-volt lantern batteries contain four 1.5-volt cells.

Integrating

Math, language arts

Science Process Skills

Observing, inferring, communicating, comparing and contrasting, formulating hypotheses, identifying and controlling variables, experimenting

How Does a Lantern Battery or Flashlight Battery Work?

(Teacher-supervised activity or teacher demonstration)

Materials Needed

- Flashlight cell (old style carbon-zinc battery) cut in half
- Paper and pencils
- Figure 8.17–1 (for comparison)

Procedure

1. Compare the cell cross-section with the materials you used in Activity 8.17.
2. Except for the galvanometer, all the types of materials used in Activity 8.17 can be identified in the battery. Can you find them?
3. Draw a picture of a battery cut in half. Label the parts. (See Figure 8.18–1.) Compare your picture with those made by other members of the group. Explain how you think it works.

Figure 8.18–1. Cross Section of Flashlight Cell

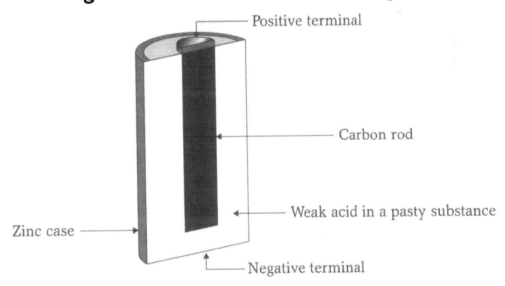

4. Have you ever seen a battery leak and damage a flashlight, a tape recorder, or some other electrical device? Batteries contain a weak acid or alkaline solution, usually in a tightly packed absorbent material. Since your battery is old, this solution has probably dried up.

5. Share your experiences and ideas with each other.

For Problem Solvers

With close supervision, dismantle a flashlight battery that is a newer, long-life variety. How is it different? Do some research and find out what materials are in this particular battery and how they react together to cause electrons to flow.

Teacher Information

Caution: This activity involves the use of cut-open flashlight cells, exposing acid substances. Although the acid involved is very weak, close supervision is very important to keep the acid from getting in eyes or on skin, or on clothing or other materials that might be damaged by the acid. It is important to wash hands thoroughly after handling these materials.

By the end of this experience, students should be able to explain in general terms how electricity is produced by chemical means. Elementary-age students should simply understand that certain materials will develop a positive or negative charge in the presence of an acid or a base (salt) and that, when these materials are connected by a conductor (wire), electricity flows through the conductor from negative to positive.

The battery shown in this activity is the lead-acid battery, which has been used for many years. Newer long-life batteries operate on the same principle, but they use different materials. You might choose to open both for comparison.

The old-style tall 1.5-volt lantern battery and the old-style lead-acid flashlight battery are constructed similar to the one shown. The 6-volt lantern battery has four such cells connected in series. Each cell produces one and one-half volts.

Technically, the word *battery* implies a collection, so the four-cell units are true batteries, while the single units are more properly called "cells." However, the single-cell unit is so commonly called a battery that to carefully distinguish between them is probably more confusing than helpful. Flashlight cells are even labeled as batteries.

Note: Batteries can be cut with a hacksaw.

Integrating

Math, language arts

Science Process Skills

Observing, inferring, comparing and contrasting, formulating hypotheses

Hands-On Physical Science Activities

How Can Electricity Be Produced by a Lemon?

(Close supervision is required)

Materials Needed

- Large, fresh lemons
- Zinc and copper strips
- Galvanometer
- Knife

Procedure

1. Just for fun, let's try to make a battery from a lemon.
2. Make two small slits in the lemon, close together, with the knife.
3. Attach the copper strip to one wire of the galvanometer and insert it into one of the slits in the lemon.
4. Push the zinc strip into the other slit. The zinc strip must not touch the copper strip, as shown in Figure 8.19-1.

Figure 8.19-1. Lemon Cell with Galvanometer

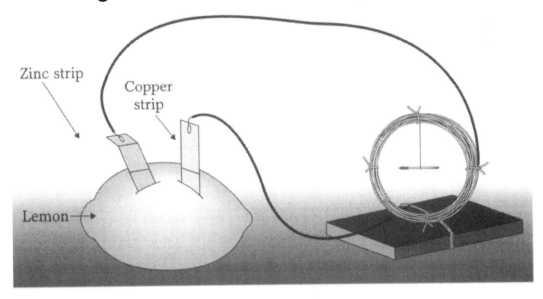

5. Wait a few seconds, then observe the needle on the galvanometer as you firmly touch the wire to the zinc strip. What happened? Wait a few seconds more and try again.

6. Using information you learned in Activities 8.17 and 8.18, explain what is happening.

7. Could you use another fruit such as a grapefruit or a tomato? Try it.

8. Discuss your observations and ideas with your group.

For Problem Solvers

There are many chemicals that can be used to create a flow of electrons. Try this activity with other citrus fruits. Try it with the juice in a cup instead of using the fruit itself. Try a penny in place of the copper strip and a silver dime (you'll have to find one that's several years old to get silver, or buy it at a coin shop) instead of the zinc. Take it from there and continue your exploration. Be sure to write everything you use and what happened, so you will remember and be able to share your information.

Teacher Information

The lemon contains citric acid, which will cause the same reaction as other weak acids. Tomatoes and citrus fruits usually contain enough acid to affect the galvanometer.

Integrating

Math, language arts

Science Process Skills

Observing, inferring, communicating, comparing and contrasting, formulating hypotheses, identifying and controlling variables, experimenting

How Can Mechanical Energy Produce Electricity?

Materials Needed

- Galvanometer
- Galvanometer coil without needle
- Extra wire
- Strong magnets

Procedure

1. Securely attach the wires from one coil to the other, as seen in Figure 8.20-1. You need enough wire to keep these coils about 2 m (2 yd.) apart from each other.

2. Hold the coil without the needle in one hand, and with the other hand move the strong magnet back and forth through the center of the coil or closely past the coil.

3. Observe the needle on the galvanometer. Can you see a relationship in what is happening?

4. Think of some words to describe the behavior of the needle. Discuss this with your teacher and other students.

Figure 8.20-1. Coil of Wire, Galvanometer, and Two Magnets

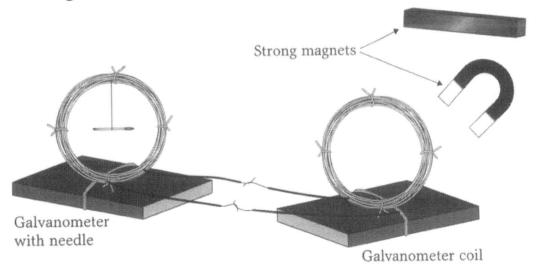

Strong magnets

Galvanometer with needle

Galvanometer coil

For Problem Solvers

Can you use the idea from this activity to design a generator? Figure out a way that you can spin a magnet within a coil of wire or spin a coil of wire within a magnetic field. Try to design your generator with more than one coil or more than one magnet.

Teacher Information

For this activity, provide plenty of wire between the two coils, so the coils can be separated by at least 2 m (2 yd.).

When electric current flows through a coil of wire, a magnetic field is created around the wire. You will remember that from your experience in making electromagnets; that's why electromagnets work. In this activity, the opposite is happening. Instead of producing a magnetic field around a coil of wire by sending an electric current through the wire, we are causing electrons to flow by moving a coil of wire within a magnetic field. Because of this phenomenon, the generator, which turns coils of wire through a magnetic field, has made electricity relatively inexpensive and plentiful. (See Figure 8.20-2.)

Observe the needle closely as the magnet goes through the coil and you will notice that it reverses the direction in which it is pointing each time the magnet moves back and forth. This shows that, as the magnetic field moves back and forth, the electrons move back and forth or alternate their direction within the wire. Electricity produced in this manner is called *alternating current.*

If you examine a small hand generator (crank type) you will notice that it is nothing more than a coil of wire turned mechanically in a magnetic field. Commercial electricity is usually produced by water or steam. A turbine, which is an enclosed wheel with curved blades, spins rapidly when water or steam is directed into it under great pressure. This provides the mechanical energy to turn huge generators. Atomic energy is sometimes used to heat water and create steam to spin the turbines.

Many people think that hydroelectric plants somehow extract electricity from water as the water rushes through the dam. Instead, the dam is engineered to direct falling water through turbines, providing the power to spin huge coils of wire within magnetic fields, stimulating the flow of electrons through the wire. See the Internet or your encyclopedia for further information.

Hands-On Physical Science Activities

Figure 8.20–2. Generator

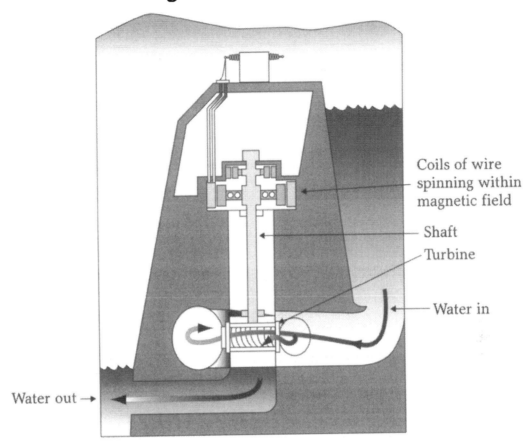

Coils of wire spinning within magnetic field

Shaft

Turbine

Water in

Water out →

Integrating

Math, language arts

Science Process Skills

Observing, inferring, communicating, formulating hypotheses, identifying and controlling variables, experimenting

How Can Sunlight Produce Electricity?

Materials Needed

- Solar cells
- Galvanometer
- Light source

Procedure

1. In recent years, scientists have been trying to find new sources of electricity to replace our rapidly diminishing fossil fuel resources (coal and oil). A most promising source is solar (sun) energy. Look at your solar cell. When light strikes this cell, a very small amount of electrical energy is produced.

2. Connect your cell to the galvanometer and shine a bright light on it. (See Figure 8.21–1.) As you turn the light on and off, observe the needle on the galvanometer. What happens to the needle?

3. Share your ideas about what causes this to happen.

Figure 8.21–1. Solar Cell in Galvanometer Circuit with Light Source

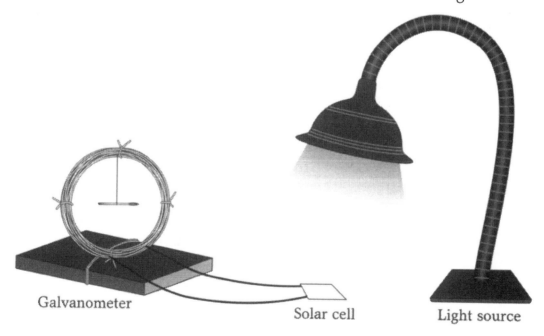

Galvanometer   Solar cell   Light source

     Hands-On Physical Science Activities

For Problem Solvers

If you can get some solar cells, learn how to connect them to operate a small light bulb. Perhaps you can find some discarded calculators and use the solar cells from them.

Do some research and see what you can learn about solar energy as a source of electricity. Make a list of all of the things you can find that are powered by solar energy. Do you think it is likely that it will ever be a major source of electricity for our homes and factories?

Can solar energy be used as a power source for electric cars? Write to automobile manufacturers and/or check with local dealers, and ask for information about current research on solar batteries and solar-powered cars. You will be able to find other sources of information on the Internet.

Teacher Information

Solar energy as a source of electricity is becoming increasingly important and common in our lives. Most children have seen solar cells used in calculators, cameras, and other devices that require small amounts of electricity.

The space program has rapidly expanded the development of this energy source. Satellites use electricity produced in this manner to recharge the batteries that provide electrical power.

A major obstacle to wide use of solar power is the limited amount of electric current each cell can produce. Huge areas of solar cells are required to produce significant amounts of electrical energy. The current produced flows in one direction, just as in flashlight batteries. Also, on the earth, solar cells as primary producers of electrical energy are limited to daylight hours and further inhibited by cloudy days. This limitation is often compensated for by using solar cells to charge batteries. The batteries are able to store an excess of electrical energy by chemical means, then provide a flow of electrons at times when the solar cells are unable to produce electrical energy.

Solar cells (often in clusters connected in series) can be obtained from many electronic supply stores.

Integrating

Social studies

Science Process Skills

Observing, inferring, classifying, communicating, comparing and contrasting, formulating hypotheses, researching

How Can Electricity Help Us Communicate?

Materials Needed

- Two small light sockets with bulbs
- Two switches
- Lantern battery
- Six 1-m (1-yd.) lengths of insulated copper wire

Procedure

1. Use your materials to construct a circuit like the one in Figure 8.22–1.
2. Press one switch. What happened?
3. Release the first switch and press the second one. What happened?
4. Press both switches at once.
5. Can you think of some use for a device like this?
6. If more wire is available, you could take one switch and light into another room.

For Problem Solvers

Look up the Morse Code in an encyclopedia or in Internet resources. If you and a friend would care to learn the Morse Code, you could have a lot of fun sending messages to each other. Or you might prefer to create your own code. Set up your telegraph sets so that you can be in separate rooms, or in separate parts of the room, and make each other's light blink with your switch.

Teacher Information

This is a variation of the telegraph. There are several ways to wire the circuit. This one is parallel. The original telegraph sets were wired in series so all keys but one had to be closed and all messages went through all the sounders in the circuit. Telegraph offices often followed the railroads and needed only one wire on the poles. The iron rails were used as the second or ground wire. The original telegraph, patented by Samuel Morse, used an electromagnet to attract a magnetic material (soft iron) and make a loud clicking sound. Telegraphers were trained to hear combinations of long and short "dots and dashes" to represent letters of the alphabet. This is called the Morse Code.

Hands-On Physical Science Activities

Figure 8.22–1. Two Telegraphs Wired to One Battery

The completion of the transcontinental telegraph near the end of the Civil War was extremely significant. For the first time, a message could travel across the nation in seconds rather than in weeks. The circuit in Figure 8.22–1 uses the electric light, which was not invented until much later.

In addition to the activity suggested in the "For Problem Solvers" section, your motivated students might enjoy designing and constructing a telegraph using electromagnets.

Integrating

Social studies

Science Process Skills

Communicating, comparing and contrasting, identifying and controlling variables

🐾 Can You Solve This Current Electricity Word Search?

Try to find the following Current Electricity terms in the grid below.
They could appear in horizontal (left to right), vertical (up or
down), or diagonal (upward or downward) position.

current	electricity	insulation
energy	electrons	positive
conductors	terminal	battery
parallel	series	circuit
switch		

```
L  P  K  J  H  Y  G  C  F  D  S  A
B  A  T  T  E  R  Y  O  E  P  W  Q
G  R  C  U  R  R  E  N  T  O  R  C
R  A  E  E  X  T  P  D  O  S  T  I
E  L  L  W  Z  E  O  U  H  I  Y  R
E  L  E  C  T  R  I  C  I  T  Y  C
N  E  C  I  M  M  T  T  M  I  U  U
E  L  T  O  N  I  E  O  U  V  I  I
R  M  R  H  W  N  R  R  Y  E  O  T
G  L  O  S  B  A  V  S  H  D  P  B
Y  I  N  S  U  L  A  T  I  O  N  N
P  O  S  E  R  I  E  S  G  F  M  I
```

Can You Create a New Current Electricity Word Search of Your Own?

Write your Current Electricity words in the grid below. Arrange them in the grid so they appear in horizontal (left to right), vertical (up or down), or diagonal (upward or downward) position. Then fill in the blank boxes with other letters. Trade your Word Search with someone else who has created one of his or her own, and see whether you can solve the new puzzle.

_____   _____   _____

_____   _____   _____

_____   _____   _____

_____   _____   _____

_____   _____   _____

Current Electricity

Answer Key for
Current Electricity Word Search

```
L  P  K  J  H  Y  G  C  F  D  S  A
B  A  T  T  E  R  Y  O  E  P  W  Q
G  R  C  U  R  R  E  N  T  O  R  C
R  A  E  E  X  T  P  D  O  S  T  I
E  L  L  W  Z  E  O  U  H  I  Y  R
E  L  E  C  T  R  I  C  I  T  Y  C
N  E  C  I  M  M  T  T  M  I  U  U
E  L  T  O  N  I  E  O  U  V  I  I
R  M  R  H  W  N  R  R  Y  E  O  T
G  L  O  S  B  A  V  S  H  D  P  B
Y  I  N  S  U  L  A  T  I  O  N  N
P  O  S  E  R  I  E  S  G  F  M  I
```

Do You Recall?

Section Eight: Current Electricity

1. What is an electrical conductor?

2. Thinking of electricity, what is an insulator?

3. What is an electrical circuit? (Consider a battery as the power source.)

4. What is a switch?

Do You Recall? *(Cont'd.)*

5. In a series circuit, what happens if one light burns out?

6. How is a parallel circuit different from a series circuit?

7. What is a resistor?

8. How does electric current affect a compass?

9. What is an advantage of electromagnets over permanent magnets?

Do You Recall? *(Cont'd.)*

10. What happens to an electromagnet when the current changes direction?

11. What is a galvanometer?

12. How do chemicals produce electricity?

13. How can electricity be generated mechanically?

14. How can sunlight be used to generate electricity?

Answer Key for Do You Recall?

Section Eight: Current Electricity

Answer	Related Activities
1. A substance through which electricity can flow easily	8.1
2. A substance through which electricity cannot flow easily	8.1
3. The path of electricity, as it moves from the battery and back to the battery	8.2
4. A device that can easily turn the power on and off	8.5
5. If one light burns out, all lights turn off.	8.6
6. If one light burns out, the other lights stay on.	8.7
7. It conducts electrical current somewhat, but resists it.	8.8
8. When electric current flows through a wire there is a magnetic field around the wire. A compass needle is a magnet and will respond within a magnetic field.	8.9, 8.10
9. Electromagnets can be turned on and off; their strength can also be changed.	8.11–8.13
10. The polarity of the electromagnet (N pole/S pole) reverses.	8.14, 8.15
11. An instrument that detects small amounts of electricity	8.16

Hands-On Physical Science Activities

Answer Key for
Do You Recall? *(Cont'd.)*

Answer   **Related Activities**

12. Certain chemicals give up electrons
    when they are in the presence of certain
    other chemicals. The resulting excess of
    electrons will sometimes flow back when
    an electrical conductor is provided.   8.17–8.19

13. When a coil of electrical wire spins
    within a magnetic field, electrons
    flow within the wire. This flow of electricity
    can be transported to other locations.   8.20

14. Sunlight can stimulate the flow of
    electrons in certain substances, providing
    power for certain applications.   8.21

Hands-On Life Science Activities For Grades K–6, Second Edition

Marvin N. Tolman, Ed.D.

Paper/300 pages   ISBN: 0-7879-7865-5

www.josseybass.com

The perfect complement to any elementary school science program, these activity books foster discovery and enhance the development of valuable learning skills through direct experience. The updated editions include an expanded "Teacher Information" section for many of the activities, enhanced user friendliness, inquiry-based models, and cooperative learning projects for the classroom. Projects use materials easily found around the classroom or home, link activities to national science standards, and include other new material. Many of the activities could become, or give rise to, science fair projects.

The study of life science at the elementary school level is best conducted through an exploration of the following topical areas: plants and seeds, animals, growing and changing animal life cycles, animal adaptation, body systems, the five senses, and health and nutrition. This book consists of more than 150 easy-to-use, hands-on activities in the following areas of sciences:

- Living Through Adaptation
- Animals
- Growing and Changing: Animal Life Cycles
- Plants and Seeds
- Body Systems
- The Five Senses
- Health and Nutrition

Current trends encourage teachers to use an activity-based program, supplemented by the use of textbooks and many other reference materials. Activities, such as those in this series, that foster hands-on discovery and enhance the development of valuable learning skills through direct experience are key to the exploration of new subject matter and to the goal of attaining mastery.

Hands-On Earth Science Activities For Grades K–6, Second Edition

Marvin N. Tolman, Ed.D.

Paper/300 pages   ISBN: 0-7879-7866-3

www.josseybass.com

The perfect complement to any elementary school science program, these activity books foster discovery and enhance the development of valuable learning skills through direct experience. The updated editions include an expanded "Teacher Information" section for many of the activities, enhanced user friendliness, inquiry-based models, and cooperative learning projects for the classroom. Projects use materials easily found around the classroom or home, link activities to national science standards, and include other new material. Many of the activities could become, or give rise to, science fair projects.

The study of earth science at the elementary school level is best conducted through an exploration of the following topical areas: air, water, weather, the earth (mapping, topography, rocks, minerals, and earthquakes), ecology, gravity and flight, and celestial bodies. This book consists of more than 160 easy-to-use, hands-on activities in the following areas of sciences:

- Air
- Water
- Weather
- The Earth
- Ecology
- Above the Earth
- Beyond the Earth

Current trends encourage teachers to use an activity-based program, supplemented by the use of textbooks and many other reference materials. Activities, such as those in this series, that foster hands-on discovery and enhance the development of valuable learning skills through direct experience are key to the exploration of new subject matter and to the goal of attaining mastery.

Dr. Art's Guide to Science: Connecting Atoms, Galaxies, and Everything in Between

Art Sussman, Ph.D.

Cloth/250 pages ISBN: 0-7879-8326-8

www.josseybass.com

Dr. Art's Guide to Science teaches major science ideas that are relevant to people's lives in ways that are very enjoyable. It expands the approach successfully used in his popular and award-winning previous book, *Dr. Art's Guide to Planet Earth*. As in that book, *Dr. Art's Guide to Science* uses a "systems thinking" framework to connect major ideas in the physical, life, and earth sciences. This book also helps the reader recognize, develop, and apply skills in using literacy strategies to enjoy and learn by reading science nonfiction.

Here's What's Inside:

- Why Science
- Two Plus Two Equals Hip-Hop
- What's The Matter?
- Energy And Dr. Art's 50th Anniversary Ball
- Forces Be With Us
- Putting The You In Universe
- Home Sweet Home
- Energy On Earth
- Life On Earth
- How Life Works
- The Famous E Word
- The Day The Dinosaurs Died
- Where Are We Going?

Art Sussman, Ph.D., is a science educator, and the author of the award-winning book, *Dr. Art's Guide to Planet Earth*. He presents "Dr. Art's Planet Earth Show" at education conferences and science centers nationally.

Other Books of Interest

Science Essentials, Elementary Level: Lessons and Activities for Test Preparation

Mark J. Handwerker, Ph.D.

Paper/320 pages   ISBN: 0-7879-7576-1

www.josseybass.com

Science Essentials, Elementary Level gives classroom teachers and science specialists a dynamic and progressive way to meet curriculum standards and competencies. Science Essentials are also available from Jossey-Bass publishers at the middle school and high school levels.

You'll find the lessons and activities at each level actively engage students in learning about the natural and technological world in which we live by encouraging them to use their senses and intuitive abilities on the road to discovery. They were developed and tested by professional science teachers who sought to give students enjoyable learning experiences while preparing them for district and statewide proficiency exams.

For easy use, the lessons and activities at the elementary school level are printed on a big 8 1/2" x 11" lay-flat format that folds flat for photocopying of over 150 student activity sheets, and are organized into four sections:

- I. Methods and Measurement Lessons & Activities
- II. Physical Science Lessons & Activities
- III. Life Science Lessons & Activities
- IV. Earth Science Lessons & Activities

Mark J. Handwerker, Ph.D., has taught science in the Los Angeles and Temecula Valley Unified School Districts. As a mentor and instructional support teacher, he has also trained scores of new teachers in the "art" of teaching science. Dr. Handwerker is the author/editor of articles in several scientific fields and the coauthor of an earth science textbook currently in use.

Janice VanCleave's Teaching the Fun of Science

Janice VanCleave

Paper/208 pages   ISBN: 0-471-19163-9

www.josseybass.com

Now you can introduce children to the wonders of science in a way that's exhilarating and lasting. In *Janice VanCleave's Teaching the Fun of Science*, the award-winning teacher and popular children's author provides key tools to help you effectively teach the physical, life, and earth and space sciences and encourage kids to become enthusiastic, independent investigators. Each science concept is presented with hands-on activities, teacher tips, key terms, and much more, including:

- Reproducible sheets of experiments and patterns
- Lists of expectations based on National Science Education Standards and Benchmarks
- Advice on preparing materials and presenting each topic
- Dozens of suggestions for extensions

As with all of Janice VanCleave's books, the format is easy to follow and the required materials are inexpensive and easy to find. With *Janice VanCleave's Teaching the Fun of Science* you can inspire, challenge, and help your students to develop a lively and lifelong interest in science.

Janice VanCleave is a former award-winning science teacher who now spends her time writing and giving science workshops. She is the author of over forty books with sales totaling over 2 million copies.